After Extinction

Center for 21st Century Studies
Richard Grusin, Series Editor

After Extinction

Richard Grusin, Editor

CENTER FOR 21ST CENTURY STUDIES

University of Minnesota Press
Minneapolis
London

An earlier version of chapter 1 was published as William E. Connolly, "Postcolonial Environmentalism, Extinction Events, and Entangled Humanism," in *Facing the Planetary: Entangled Humanism and the Politics of Swarming* (Durham, N.C.: Duke University Press), 151–74; all rights reserved; reprinted by permission of the copyright holder. An earlier version of chapter 2 was published as Jussi Parikka, "Planetary Goodbyes: Post-History and Future Memories of an Ecological Past," in *Memory in Motion: Archives, Technology, and the Social,* ed. Ina Blom, Trond Lundemo, and Eivind Røssaak (Amsterdam: Amsterdam University Press, 2016), 129–52. An earlier version of chapter 4 was published as Joseph Masco, "The Six Extinctions: Visualizing Planetary Ecological Crisis Today," *Environmental Philosophy* 14, no. 1 (2017), doi:10.5840/envirophil201611741. An earlier version of chapter 5 was published as Cary Wolfe, "Condors at the End of the World," in *You Must Carry Me Now: The Cultural Lives of Endangered Species,* by Bryndís Snæbjörnsdóttir and Mark Wilson, ed. Ron Broglio (Styrsö, Sweden: Förlaget 284, 2016).

Published by the University of Minnesota Press
111 Third Avenue South, Suite 290
Minneapolis, MN 55401-2520
http://www.upress.umn.edu

ISBN 978-1-5179-0289-6 (pb)
ISBN 978-1-5179-0288-9 (hc)

A Cataloging-in-Publication record for this book is available from the Library of Congress.

Printed in the United States of America on acid-free paper

The University of Minnesota is an equal-opportunity educator and employer.

25 24 23 22 21 20 19 18   10 9 8 7 6 5 4 3 2 1

Contents

Introduction

Richard Grusin

This volume marks the conclusion of an informal trilogy of books that culminates my initial five-year tenure as director of the Center for 21st Century Studies (C21) at the University of Wisconsin–Milwaukee. Like the preceding two volumes—*The Nonhuman Turn* and *Anthropocene Feminism*—this one originated in the annual spring conference sponsored by C21.[1] Each of the three volumes responds in different but overlapping ways to the question I posed at the beginning of my directorship in 2010: "What is twenty-first-century studies?" Although they offer only an incomplete snapshot of this complex area of interdisciplinary research, the three volumes speak to one another in interesting and productive ways.

After Extinction in particular picks up on the speculative spirit of the previous year's conference (and subsequent volume) on anthropocene feminism, which began as a kind of experiment in collaborative theorizing. Like that conference, the conference on which this volume is based started with a single question: what comes after extinction? In ongoing debates about the Anthropocene, this question has been very much present, in relation both to the anthropogenic sixth extinction and to the premediated disappearance of humans as a species. Indeed, to periodize the Anthropocene is already to assume a future world in which human presence on Earth has been reduced to a lithic layer. The Anthropocene contains within it both the anticipation of human extinction and the imagination of how such extinction would manifest in Earth's crust. Our predominant understanding of the extinction event under way in the twenty-first century relates to natural species extinctions caused largely by human actions. But as our conference and volume on the nonhuman turn in twenty-first century studies have suggested, categorical distinctions between humans and nonhumans or culture and nature are no longer tenable—if they ever really were. Indeed, as Darwin was not even the first to note, mass extinction events preceded the appearance of humans on the planet.

The concept of species extinction first emerged in the eighteenth century to account for the discovery of fossils that had no living correlates. Those fossils challenged predominant Christian notions of the Great Chain of Being in which Nature was understood as a complete and perfect whole, created by God without gap or imperfection. If Nature contained all and only those species that were divinely created, how could any of them be allowed to go extinct, or how could new species emerge? Darwin's revolutionary mid-nineteenth-century theory of natural selection treated the origin of species as reliant on chance and accident, not divine purpose, and therefore offered a fundamentally different understanding of extinction not as an aberration but as part and parcel of the process of natural selection. In addition to the ongoing processes of species creation and extinction, scientists have now come to agree that the planet has previously undergone five mass extinctions and that we find ourselves today in the midst of a sixth. Unlike these others, the sixth extinction is attributed to the agency of humans, or more accurately to the agency of nonhuman processes set in motion by humans, particularly the technologies of industrial capitalism and globalization that continue to operate both in service to and irrespective of human intent and purpose.

In addition to the mass extinction of species, today we think as well of the extinction of cultural forms: languages, customs, and traditions; craft and artisanal skills; media, technologies, and operating systems; public institutions. As Félix Guattari wrote in *Three Ecologies,* "it is not only species that are becoming extinct but also the words, phrases, and gestures of human solidarity."[2] In the face of this extended sense of extinction, asking what comes after extinction is not only to inquire about the future of humans and nonhumans but also to investigate to what extent the concept's theological origins still inflect current understandings of extinction. Does the very idea of extinction bear traces of an ontology that is alien to natural, social, and human scientists in the twenty-first century?

To ask what comes after extinction, then, is on one hand to refer to the event of extinction. What comes after such events, whether local events like the extinction of a species or more massive events like the much anticipated sixth extinction? If we think of such events not as destructive or final but as generative, as what philosopher Alfred North Whitehead would call "occasions of experience," then what comes after these occasions? What comes next?[3] Is extinction something that only happens belatedly, after there are already species or forms or practices in place? Or does the very possibility of extinction work in a more radical form, as already present in the origin of species more generally? Is there a sense in which extinction might be prior to or even generative of the evolution or emergence of any form of life or being? In addition to the event of extinction, we meant also to refer to extinction as a concept. What comes after thinking extinction? Where are we left after we are placed in the position, individually and collectively, of having to think about endings and what comes after them in the deepest sense of the term? What happens to writing, theory, and philosophy after thinking the concept of extinction?

Extinction is not simply an absence but a geophysical event that occurs in space and time. In soliciting others to think "after extinction" with us, we were also interested in exploring some of the different meanings of *after*. In the predominant temporal meaning of the word, what follows upon extinction? Are there historical models or examples of what comes next? Can these past extinctions measure up to present-day events, or do the possible scales on which extinction might operate today make such comparisons incompatible? Is extinction terminal, or can species return, à la *Jurassic Park* or European projects to restore the aurochs or Przewalski's horse? Can dead or dying languages be revitalized?

The invitation to think after extinction also invokes the question of what will remain or endure spatially or physically after the next mass extinction. How can one try to act "at a distance" after extinction to plan for, prevent, or preempt the end of crucial life-forms—by establishing seed banks, for example, or stockpiling DNA? How does the extinction of one species threaten the lifeblood of the entire biosphere (for example, the impact of bee colony collapse on particular flora and fauna as well as on human practices like agriculture)? Have new artifacts surfaced either as sentinels or fossils of extinction (for example, animal carcasses washed up on shore filled with plastic or mutant plants in irradiated nuclear test fields)? Even if extinction has always been thought of as impacting a larger ecology, has the *scale* of risk changed in light of the accelerated planetary and even interplanetary sociotechnical networks of the twenty-first century?

Finally, we were interested in the meaning of *after* that points toward resemblance or mimesis. What would it mean for an image, graphic, text, video, or film to "take after" the concept of extinction, to mediate it in such a way as to resemble or be mimetic of extinction? What would it mean to be "after extinction" in the sense that a painting is "after O'Keeffe" or a child "takes after" its parent? To be recognized as coming after extinction, an event or occasion must be seen as being related to extinction, to have been consequent or emergent from it. Thus we were interested in exploring how future extinctions were being premediated in a variety of formal and informal, print, audiovisual, and networked media. What forms of knowledge (or nonknowledge) emerge from such anticipatory pursuits?

The answer to this last set of questions points toward my own work on premediation, which names the logic with which one tries to mediate future events, a logic that has intensified in the twenty-first century. Such premediation, as I argued in relation to the run-up to the Iraq War, terrorist threats, and other potential catastrophes, serves simultaneously to generate anxiety about future catastrophic events and to provide reassurance that they have already been anticipated, remediated, and survived. Could it be possible, then, that our current preoccupation with questions of extinction, like our preoccupation with the Anthropocene in all of its varieties, represents not an engagement with the pressing concerns of the twenty-first century but rather the opposite? Could our theorization and speculation about anthropogenic mass extinction be a way of escaping, avoiding, or minimizing such concerns through the premediation of anthropogenic, climatological apocalypse? Or has premediation by the United States and other developed nations of strategies and plans to survive climatological disaster already brought into existence catastrophes like the sixth mass extinction, making them already integral both to the present and future of humans and nonhumans on the planet?

The authors in this volume address some of these questions as well as posing some of their own in the chapters that follow. Several authors look at the question of premediation through analysis of some of the different ways that contemporary artists and writers have tried to represent extinction through the creation of works which are, in the mimetic or representational sense, "after extinction." Others look at the interrelation of histories of species extinction with histories of racism, colonialism, and imperialism. The question of de-extinction is also taken up by a couple of the authors, and interwoven in different ways into all of the chapters is the question of how we might best embrace a future after the concept of an impending sixth mass extinction has become an accepted feature of our ecological and political landscapes.

In "Extinction Events and Entangled Humanism," William E. Connolly offers arguably the volume's most optimistic approach to the question of how to live in the face of impending climatological and environmental disaster. As the title of his contribution implies, he argues for what he calls an "entangled humanism," which emerges from pluralist accounts of hopefulness after extinction, drawn around some of the major thinkers of nineteenth-century Western intellectual history. Connolly challenges the idea that the humanist exceptionalism of the Western tradition can stave off the nihilisms prompted by impending ecological disaster, asking whether such exceptionalism might "run the grave risk of becoming a new mode of nihilism during the era of the Anthropocene." To stave off this nihilism, Connolly does not want to reject humanism but to "transfigure humanist exceptionalism into affirmation of entangled humanism in a fragile world." Such entangled humanism must be multiscalar, beginning with the acknowledgment of the "bacterial microagencies" of the human body and moving through our prehuman physiological and genetic inheritance to the "interspecies symbiosis," "racial and regional struggles," and "planetary, partially self-organizing processes" that "impinge upon every other entanglement." To combat the "passive nihilism" that such entanglements threaten to invoke, Connolly offers an optimistic call "to give the politics of swarming and cross-regional general strikes a try," aiming first at those extractive and consumptive regimes that have caused the greatest planetary damage but recognizing that "if that urgently needed improbability proves to be impossible, it may be time to move to the next item on the agenda."

Where Connolly tries to imagine strategies for acting today to prevent a future extinction toward which we seem to be inexorably heading, Jussi Parikka tries in "Planetary Memories: After Extinction, the Imagined Future" to imagine what the present might look like after extinction, as seen through the eyes of future or posthistory. Parikka takes as his two central examples Finnish designer, philosopher, and artist Erkki Kurenniemi's imagining of a postplanetary human future, "In 2048," and the best-selling 2014 book by Naomi Oreskes and Erik Conway, *The Collapse of Western Civilization,* narrated from the end of the twenty-first century by a future historian from China. Parikka is concerned with what he calls the "politics of chronoscapes," which underscores the "multiple historical and temporal ecologies" that we face today. Drawing on Wendy Brown and Sarah Sharma, Parikka urges the importance "of thinking about the contemporaneity of the present as informed by multiple temporalities and synchronization across the time scales." Not unlike Connolly's call for an entangled humanism, Parikka's call for a politics of chronoscapes insists upon entangled temporalities, urging that we "think of the uneven and multiple overlapping temporalities that help to determine the otherwise broad concepts of *the political contemporary.*" Parikka finds in Vilém Flusser's "idea of posthistory" a promising tool to imagine what the future might look like after extinction.

In "Photography after Extinction," Joanna Zylinska explicitly takes up the "exhortation" in the C21 conference call for papers to "think of the event of extinction not as destructive or final, but as generative." Not unlike Parikka or Connolly, she seeks "to consider what happens to writing, theory, and philosophy—but also to art and photography—as devices for enabling a radically different set of arrangements for the world (aka a radically different politics) *after thinking the event of extinction.*" Through a fascinating reading of late-nineteenth-century English scientist William Jerome Harrison, Zylinska brings out the nonhuman aspects of photography's historical dependence on the sun as a way to introduce four different examples of contemporary photographic artists who deal in different but striking ways with imagining a world after extinction. Hiroshi Sugimoto's exhibition *Lost Human Genetic Archive* pursues the connections between fossilization and photography in several "just after extinction" scenarios. In *Club Disminución,* Alexa Horochowski juxtaposes in one work a fossil of a trilobite with one of a credit card to situate us in a posthistory well after the extinction of the human race. Penelope Umbrico's works focus on the remediation of the sun and sunsets by media technologies like Flickr or the pixelated digital screen. And Edward Burtynsky's photographic series *Oil* "features large-scale images of oil fields in Azerbaijan, the United States, and Canada; discarded or burning tire piles in California; and oil refineries"; these beautiful, "predominantly bird's-eye-view images" dramatize the risk of "aesthetics acting as an anesthetic against the urgency of the environmental situation." For Zylinska, each of these works exemplifies the role photography can play in "envisaging a new energetics," in which we recognize its and our dependence on "oil, a light distilled from death."

Like Zylinska, Joseph Masco takes up "after extinction" in relation to representation or mediation. In "The Six Extinctions: Visualizing Planetary Ecological Crisis Today," Masco examines the complexity of ways in which ecological crisis has been envisioned in contemporary art, contrasting it with the more singular image of the mushroom cloud, for example, as a visualization of the nuclear threat. Like Parikka and Zylinska, Masco seeks to make sense of the complex temporalities of extinction: "This chapter is therefore ultimately about conceptualization, about how to think on temporal and spatial scales that exceed human senses." The heart of Masco's chapter is a detailed survey of Hamza Walker's *Suicide Narcissus,* a 2013 exhibition at the Renaissance Society of Art in Chicago. Walker's exhibition, Masco writes, is "an explicit engagement with the aesthetic pull of extinction," which "helps us think about the limits of human perception as well as the psychosocial effects of radical collective endangerment." The six works include Lucy Skaer's *Leviathan's Edge,* an installation featuring a whale skeleton; Katie Paterson's *All the Dead Stars,* a laser-etched image on anodized aluminum; Thomas Bauman's *Tau Sling,* a mechanized installation of wood, rope, and mirror; Daniel Steegmann Mangrané's film *16mm*; Nicole Six and Paul Petritsch's video *Spatial Intervention I*; and Haris Epaminonda and Daniel Gustave Cramer's *The Infinite Library,* a dozen of a series of sixty artist's books. As

Masco engagingly demonstrates through his reading of the exhibition, "the immediate answer to the problem of visualizing planetary ecological crisis today is not to consolidate climate change into a single image, offering a mushroom cloud for a new emergency, but rather to proliferate modes of conceptualization and visualization of ecological conditions that can allow wide contemplation of the complexity of human interventions into natural processes and, most importantly, evolve radically with those understandings."

Cary Wolfe also takes up the problem of how to visualize extinction, focusing in "Condors at the End of the World" on a photographic series from the exhibition *Trout Fishing in America and Other stories,* staged by artists Bryndís Snæbjörnsdóttir and Mark Wilson at the Arizona State University Art Museum in 2014–15. The series Wolfe focuses on consists of "photographs of the frozen, preserved bodies of fourteen dead California condors, each printed above a transcribed text about the bird taken from conversations with the biologists working in the conservation program." Although the California condor became extinct in the wild in 1987, conservation biologists have been engaged in a successful program reintroducing them to the wild in the Grand Canyon area. The frozen birds photographed for the exhibition had been found dead in the wild, often "due to lead poisoning from feeding on animals killed by hunters with lead bullets." Wolfe puts the exhibition in dialogue with the second volume of Jacques Derrida's *Beast and the Sovereign* seminars, in the process of which Wolfe deploys Derrida to work through questions of death, archive, alterity, and the animal–human distinction. Doing so leads him to consider extinction as something like the Derridean supplement, both excessive to nature and necessary to complete it. Wolfe concludes with the claim that "there is nothing more 'natural' than extinction—it is an event that happens '*there,*' in nature, and *has* happened with the vast majority of species that have ever existed; but at the same time, extinction is and can never be a 'natural' event because it always takes place within an horizon of 'world' and its governing principles—including, of course, the principle of 'biodiversity'—that *we* create through 'stabilizing apparatuses.'"

The remaining chapters shift the volume's focus from the aesthetic and philosophical aspects of thinking after extinction to the political origins and consequences of acting after extinction. Nicholas Mirzoeff and Claire Colebrook each take up the connections between the logic of extinction and those of racism, colonialism, and imperialism. Mirzoeff's chapter is designed explicitly as a "provocation and an opening to a broader discussion." The broader discussion he seeks to initiate is one that takes up his proposition "that the very concept of observable breaks between geological eras in general and the definition of the Anthropocene in particular is inextricably intermingled with the belief in distinct races of humanity in the former instance and the practices of (neo)colonialism in the latter." Mirzoeff challenges the "substantial body of humanities scholarship being produced in response to the combined impact of the Anthropocene turn, the material turn, and the nonhuman turn" for generating "a turn away from understandings of race, white supremacy, colonialism, and imperialism."[4] With an eye toward countering such a turn, Mirzoeff provides vignettes from four eighteenth- and nineteenth-century naturalists to unfold the mutual reinforcement of racial and geological breaks in the drawing of lines between epochs and races. He then turns to current debates on where to locate the "golden spike" to mark the beginning of the Anthropocene. These debates demonstrate the racialized nature of the concept of the Anthropocene itself, insofar as the various candidates constitute the choice as one "not just for humans over the nonhuman" but largely between "a certain highly privileged group of humans over all other humans and the nonhuman." Mirzoeff concludes that "the formation of a discourse of extinction was an entanglement of hypervision and description designed to negotiate and negate the era of abolition and revolution by generating lines of force that separated and distinguished permanent races in the human and nonhuman worlds."

Claire Colebrook also argues for the interrelationship between extinction and race, particularly through the discourses of universalism and colonialism. She suggestively links the concepts of extinction and disability to think through the question of what comes after extinction. Rather than take disability or extinction as secondary or derivative conditions of personhood, Colebrook argues "that the problem of disability runs to the very heart of the extinction-logic that enables the political tradition of the person." Taking up philosophical discussions of "the life worth living," which attempt to calculate which (human or nonhuman) lives should be valued or preserved over others, she argues that "when philosophers dispute about a life worth living, arguing whether a life is able enough to live, they are part of the same voice that can observe fragments of the human species as a genus, or a particular kind of a general species, over which a single voice might range." By valuing human rational thought as the measure of independence and capability, without realizing that its dependence on "networks of labor and technology" depend on the capacities of others whose lives were generated as "unworthy and dis-abled," such philosophers engage in an offensive against racialized others. Consequently, Colebrook argues, "any epoch of thriving and fecundity takes place at the expense of some lives," and "all ages are ages of extinction. What makes our time—the sixth mass extinction—more *intense* is that questions that have always haunted political personhood are now becoming more explicit." Her essay concludes with a case study of the "last Tasmanian Aborigine" as an example of the racialization of extinction or genocide. "Asking the question of the good life, of how 'one' ought to live, is both genocidal and extinction-generating." In a claim that harmonizes with Mirzoeff's argument, Colebrook avers that "this seemingly intractable and universal problem is a problem for a portion of humanity, and a portion that has the logic of extinction at its heart."

Where Colebrook and Mirzoeff relate the logic of extinction chiefly to colonialism and racism, Ashley Dawson adds to this focus an exploration of the ideological work done by advocates of de-extinction to perpetuate the destructive processes of capitalism that have brought us to the brink of the sixth mass extinction in the first place. For Dawson, "the extinction crisis is at once an environmental issue *and* a social justice issue, one that is linked to long histories of capitalist domination over specific people, animals, and plants." Framing his chapter against the idea that the cause of the sixth mass extinction, like the Anthropocene, is "divined in some general capacity of human beings to destroy the natural world," he argues that "extinction needs to be seen, along with climate change, as the leading edge of contemporary capitalism's contradictions." He shows how de-extinction's goal to preserve biodiversity "dovetails perfectly with biocapitalism," particularly how they both believe in "a series of catastrophic crises that ultimately generate new forms of complexity." Dawson traces the support for many de-extinction and biogenesis efforts to "ecomodernism," a movement rooted in "the California ideology" of Stewart Brand and other Silicon Valley intelligentsia that seeks to engineer a way out of the ongoing climate crisis and into a "good Anthropocene." While "Brand and his collaborators characterize themselves as heretics bringing dissident and fresh perspectives to a hidebound environmental movement," Dawson argues that "their arguments are completely consistent with the central myth of global capitalist culture: the belief in the possibility of unending growth." He concludes his chapter with a discussion of the genocidal impacts of this "New Extractivism" in the Brazilian Amazon basin, where "more than sixty big dams and other infrastructural projects are currently slated" in an effort to develop these untapped resources for global biocapitalism. Similar to Mirzoeff, Dawson calls for "an anticapitalist movement against extinction," one that would "reject capitalist biopiracy and imperialist enclosure of the global commons, particularly when they clothe themselves in arguments about preserving biodiversity."

In the volume's closing chapter, Daryl Baldwin, Margaret Noodin, and Bernard C. Perley explore the complex relations between the de-extinction of native languages and the processes of colonialism and imperialism that have led to the extinction not only of native languages but also of native peoples and cultures. Where Dawson shows the complicity of de-extinction movements with the ideology of biocapitalism, Baldwin, Noodin, and Perley show how the very concept of extinction is part of white colonialist, not native, epistemologies. Among the important points made in this chapter is that not only are "indigenous languages at risk of extinction" but so also are "the social relations that are mediated by those languages." Furthermore, the authors point out, the very idea that a language could be considered to go extinct, as the international linguistic community now does, is a concept that is foreign to the three native languages represented among the authors. Extinction, they write, "is not only a foreign concept but an invasive one," borrowed from biology. To "challenge Western notions of extinction and endangerment," the chapter offers three case studies. Because the linguistic structure of the Myaamia language does not allow for the idea that language could go extinct, Daryl Baldwin and his collaborators "exercised Myaamia linguistic sovereignty by changing the metaphor from *extinct* to *sleeping*" to create "the possibility that the Myaamia language can be awakened." Baldwin has been deeply involved in the project called Myaamiaki Eemamwiciki: Miami Awakening. Bernard Perley is engaged in a project to challenge the criteria by which language endangerment is evaluated, showing, for example, how "three popular linguistic and world heritage sites present three different assessments of the relative vitality of the Maliseet language." Describing several ongoing efforts to revitalize Maliseet linguistic interactions, the authors explain their "commitment to integrating all social relations in everyday interactions to assure creative and innovative futures for language, culture, and identity while drawing from the foundational substrate of tradition and continuity." Finally, the authors describe how Margaret Noodin's efforts to support the teaching of Anishinaabemowin from elementary school to universities help to make evident the prevalence of Anishinaabe languages, which are currently spoken "in more than two hundred Anishinaabe communities in Quebec, Ontario, Manitoba, Saskatchewan, Alberta, North Dakota, Michigan, Wisconsin, and Minnesota." Because of this wide range, its "subtle connectivity is not obvious to external observers. For example, Anishinaabemowin is not one of the 2,467 languages mentioned in the UNESCO Atlas of the World's Languages in Danger," even though its "most common variant, Ojibwe, is noted in four locations as vulnerable, definitely endangered, and severely endangered" and other variants are either underreported or not listed at all.

Although these "are only three of many American Indian communities who have survived the first round of human extinction," Baldwin, Noodin, and Perley conclude their chapter and the volume itself with the hope that "as we all anticipate and face the next round of mass extinctions, we can look to American Indian strategies of awakening, emergent vitality, and sovereignty for guidance in surviving what will become a global New World." On behalf of all of the contributors to this collection, I feel confident in saying that we welcome such guidance as we move forward after extinction.

NOTES

1. Richard Grusin, ed., *The Nonhuman Turn* (Minneapolis: University of Minnesota Press, 2015), and Grusin, ed., *Anthropocene Feminism* (Minneapolis: University of Minnesota Press, 2017).

2. Félix Guattari, *The Three Ecologies,* trans. Ian Pindar and Paul Sutton (London: Athlone Press, 2000), 30.

3. Alfred North Whitehead, *Process and Reality* (New York: Free Press, 1978).

4. Not coincidentally, this body of scholarship is represented in all three volumes of my edited trilogy in twenty-first-century studies. In the introduction to *The Nonhuman Turn,* I address the concern that a turn to the nonhuman represents a turn away from liberatory politics: "Considered more broadly, the nonhuman turn often invokes resistance or opposition from participants in liberatory scholarly projects—for example, feminist critiques of sexual violence, critical race studies, or holocaust studies—which work precisely against the objectification of the human, its transformation into a nonhuman object or thing that can be bought and sold, ordered to work and punished, incarcerated and even killed. For scholars working on such politically liberatory projects, who have labored so hard to rescue or protect the human from dehumanization or objectification, the nonhuman turn can seem regressive, reactionary, or worse, particularly if it is identified solely with the turn to objects as fundamental elements of ontology. Motivated partly by social constructivism, many practitioners of politically liberatory scholarship share the belief that any appeal to nature, for example, as possessing causality or agential force, could only operate in service of a defense of the status quo rather than fostering a more capacious sense of becoming and construction than social constructivism imagined. But this does not have to be the case. A concern with the nonhuman can and must be brought to bear on any projects for creating a more just society" (xviii).

1

Extinction Events and Entangled Humanism

William E. Connolly

Human exceptionalism, consummate knowledge in principle, capitalist and communist mastery over nature, belonging to a beneficent world, cultural internalism, sociocentrism—all these contending world pictures demand revision today. What is wrong with them?[1]

Let us consider a couple of examples briefly. Karl Marx, in his masterful attempt to reveal the contradictions of capitalism, bypassed the insights his early flirtation with swerves in nature in Epicurean thought might have provided. He, after the early alienation essays organized around an organic model of species life, basically treated nature as a deposit of resources for human extraction and production that could be exploited more rationally, abundantly, and equitably in a communist society. Of course, there are subterranean movements in his thought that could still be mined today. And they must be. But that was the dominant trend. He bought into the then current notion of *gradualism* of nonhuman processes and the susceptibility of many nonhuman processes to human mastery. In that sense, he was a follower of the early Darwin of *On the Origin of Species,* whose theory of evolution enchanted him.[2]

Max Weber, whose brilliant study of the origins (or at least consolidation) of industrial capitalism in northern Europe remains critical to contemporary understanding, insisted that Calvinism played a key role in consolidation of the investment, managerial, worker, and consumption practices needed during a period of capitalist accumulation. In that respect, he trumped in advance contemporary secular theories that tacitly insist that the objects of social explanation must themselves be wholly secular so that the investigators will not have to slide into the slippery topics of religion, metaphysics, and ontology. To many sociocentrists, the latter zones are peripheral to the serious business of explanation in political science, economics, and sociology. And since nonhuman planetary processes such as climate, species evolution, ocean currents, glacier flows, and water self-filtration processes are assumed to be set on long, slow time—gradualism—they are mostly relevant as part of the environment or objects of production. Even many who have emphasized the role capitalism has played in creating radical climate change tacitly act as if these nonhuman processes were set on a gradual trajectory before the advent of capitalism. They were not. That assumption is the second-order expression of sociocentrism.

Weber cracked that mold a bit with his exploration of resonances between Calvinism and early industrial capitalism, indicating how disciplines flowing from cultural orientations to an extrahuman divinity played an active role in capitalism. Nonetheless, he assumed the other nonhuman processes noted above to form the settled background of human activities. Sure, industrial capitalism could *intervene* in nature until, finally, it uses up the last ton of coal upon which it depends.[3] An important early insight. *But for him as for Marx, this is an intervention in processes that in themselves move on long, slow time.* Sociocentrism in the human sciences demands a gradualist image of planetary processes to be; in so doing, it underplays multiple, interacting, nonlinear, nonhuman processes with autocatalytic capacities that may remain stable for a while and then shift rapidly.

It would be easy to add economists such as Friedrich Hayek, Milton Friedman, and John Maynard Keynes to this list, though Karl Polanyi, among others, represents a partial break with this tendency. Hayek, for example, noted the role of self-organization in the evolution of species and language as well as markets, but he neither brought the periodic volatilities in social movements to bear on these processes (they interfere with market processes) nor did he ask how attention to active nonhuman fields would disrupt his magical equation between market self-organization and impersonal rationality.[4]

Cultural internalism and human exceptionalism in the humanities follow complementary trajectories. These tendencies are discernible in the work of Hannah Arendt, Walter Benn Michaels, George Kateb, Jurgen Habermas, John Rawls, and innumerable others.[5] Cultural internalists worry, properly so, about how practices of reductionism in the natural sciences, when accepted and applied as behavioral disciplines, squeeze space for human agency, dignity, freedom, and political enactment. But they then respond to these dangers by turning away from alternative work in philosophy, neuroscience, and complexity theory that, first, refashion our understandings of human agency without adopting either an exceptionalist or reductionist route; second, identify modes of microagency within and beyond the human estate that help to constitute and limit human agency; and third, as a corollary, identify large, nonhuman, partially self-organizing processes that enter into fortunate or fateful conjunctions with capitalism, socialism, democracy, and freedom. The internalists also repress the temptation to pour an aesthetic element into nonhuman processes themselves, though Darwin himself did so in 1871 when he modified an early version of his theory and talked about the role of intraspecies attractions in evolution. The exceptionalists ignore such relational processes to monopolize agency, dignity, meaning, and artistry for the human estate alone. They thus seek to carve out a precarious space for human exceptionalism even as the stances they stake out increasingly reduce the effective space of the agency they can identify. Each new advance in neuroscience and biology is received as a threat to them rather than also being a site at which a mass of microagents can be identified that contribute to human agency. Why? We *must* be the only artists, poets, agents, and bearers of meaning on the face of the earth, they insist.

Some reductionists in the natural sciences emerge as twins of human exceptionalism. They, too, sometimes postulate an "anthropic exception" that enables human beings to explain the rest of the world in a way that lifts them (inexplicably) above modes of nature to be explained in reductive terms.[6]

This story about sociocentrism, exceptionalism, cultural internalism, and scientific reductionism could go on (and on). It does, indeed, need to be given much more nuance and qualification, since we are talking about *differential propensities* here rather than uniform commonalities. Nonetheless, let's turn to contemporary work in geology, paleontology, and species evolution that calls the background assumptions of these stories into question.

EXTINCTION EVENTS AND BUMPY TEMPORALITIES

Charles Lyell and Charles Darwin, respectively the great nineteenth-century geologist and theorist of species evolution, were both gradualists about geological and species change. Living between 1797 and 1878, Lyell became a preeminent geologist. Geological strata change very slowly, he said. The "wild" theories of one predecessor, Georges Cuvier, about a series of periodic "catastrophes" in nature led by the extinction of dinosaurs were to be rejected in the name of respecting the everyday grind of normal science over unfounded speculation. Darwin's own theory of natural selection and survival of the fittest underwent significant revision in 1871, with the publication of *The Descent of Man.*[7] There he added aesthetic and sexual elements to the theory for the evolution of species with vertebrae, a *teleodynamic* amendment inserted into his earlier theory that calls the *neo*-Darwinism vintage into question even before the very formation of the latter theory. But, in doing so, he nonetheless retained a gradualist outlook. He claimed, for instance, that the residues of bodily hair, the curtailed ability to smell, the capacity of some of us to wiggle our ears (I can), and the shape of our organs of locomotion express shifts over millennia from earlier formations that gradually allowed the evolution of humanity to occur. The rudiments connect us to a long evolutionary tree linked to slow geological change. Gradualism and evolutionary progress involved extinctions, but they seem to be tied to an overall advance in the complexity of species at the top of the pyramid.

Indeed, the Lyell–Darwin conjunction created one version of racism and white supremacy, while the Cuvier–Agassiz theories of species eternalism engendered another, with the latter's hierarchy among the eternal forms. The Lyell–Darwin version construes "Caucasians" to be at the pinnacle of the "evolutionary tree," with other "races" lower on the scale, an assumption with horrendous consequences for the perpetuation of racism.

The desire to avoid both perils probably played a role in the formation of sociocentrism in the human sciences, though that response can no longer suffice during the era of rapid climate change and acute awareness of other nonhuman force fields with self-organizing capacities.

Early critics of gradualism, including Henri Bergson, pounded away at the problem of how the evolution of one organ *could* be gradual and unilateral unless it was intimately involved with a set of other shifts in organs with which it was also intimately involved. They thought about more rapid changes in entire organisms.[8] But Bergson's objections were generally thought to point merely to sore spots in the dominant theory that would eventually be ironed out within a gradualist image of the evolutionary tree. Better to retain the sore points than to insinuate the divine element of élan vital into a bumpy and creative evolutionary process. The identification of *symbiogenesis* by Lynn Margulis and the introduction of theories of *epigenesis* later helped to challenge such neo-Darwinist presumptions, though that struggle is still very much in play.

After Darwin and Lyell, critiques of gradualism surfaced only sporadically until as late as the 1980s. Then an essay appeared by Luis Alvarez, a physicist, about that huge asteroid in Mexico that wiped out dinosaurs and several other species rapidly sixty-five million years ago.[9] About 50 percent of life disappeared during a short period. The first shock to gradualism. After a period of intense debate, the rapid extinction of dinosaurs became the established view. Here was a mass extinction event over a short period followed by a turn in the very trajectory of evolution. If several such instances were discovered, species evolution would now look more like a series of interacting, bumpy temporalities bushing out in several directions than a single evolutionary tree set on a gradualist trajectory of ever-growing complexity. More like a rhizome than a tree.

And new examples arrived quickly. There was the near-extinction of life as such 250 million years ago, when up to 90 percent of life succumbed over perhaps a period of twenty thousand to two hundred thousand years. There was another major extinction 450 million years ago and another yet around 200 million years ago. Human beings faced a trauma from a monster volcano seventy-five thousand years ago. And there was also the final extinction of Neanderthals around twenty-eight thousand years ago. Today we are participating in a new mass, cross-species extinction event, triggered by extractive capitalism, a new era of climate change entangled with it, the loss of evolutionary niches for a variety of species, and the impoverishment and environmental degradation of several regions that are treated as sites of extraction and dumping grounds for toxic wastes produced elsewhere. Studies by Rob Nixon and Anna Tsing reveal dramatically how the lines of separation constructed in the 1980s and 1990s between postcolonial theory and Euro-American, middle-class environmentalism can no longer be allowed to stand.[10] My view is that the recent line of cross-territorial reflection pursued by these authors is invaluable, and it too needs to be augmented along another dimension. We need to augment both trajectories by engaging species extinction events in the past that help to reveal things about what may be happening today at the planetary level. Let's look more closely at just two such events, then: the great extinction event 250 million years ago and the more recent decimation of our kissing cousins, the Neanderthals. I tiptoe into this literature as an amateur entering a domain of inquiry that is both unsettled and highly pertinent to the human sciences and humanities.

TWO EXTINCTION EVENTS

Two hundred fifty-one million years ago, up to 90 percent of life on Earth was extinguished over a short period in geological time. The causes of this devastation are still being debated, and indeed, it only became an object of close exploration recently. Hypotheses of another major asteroid impact have been advanced but seem to be based on scant evidence. One scenario I will review links the event to a series of methane bursts that occurred after a rapid period of global warming. The warming period had itself been caused by volcanic eruptions in Siberia—the Siberian Traps—releasing huge amounts of basalt and other gases into the air. This event alone, however, did not suffice to cause the extinction. The "repeated eruptions," however, may have sufficed to warm the air over the Antarctic Ocean enough to lift the ice cover over methane fields there. As Michael Benton contends, "release of carbon dioxide from the eruption of the

Siberian traps led to a rise of global temperatures of 6 degrees Celsius or so [about 11 degrees Fahrenheit]. Cool polar regions became warm and frozen tundra became unfrozen. The melting might have penetrated to the frozen gas hydrate reservoirs located around the polar oceans, and massive volumes of methane may have burst to the surface of the oceans in huge bubbles. This further input of carbon into the atmosphere caused more warming. . . . So the process went on, running faster and faster. The natural systems that normally reduce carbon dioxide levels could not operate, and eventually the system spiraled out of control, with the biggest crash in the history of life."[11]

As the *may*s in the preceding quotation suggest, this remains a speculative theory, though there is evidence to support it, and the event itself now seems undeniable. The evidence of a double whammy followed by a series of self-amplifying processes is stronger than that so far offered in support of other theories. Moreover, there is solid evidence of such a methane burst fifty-five million years ago, with severe effects of its own.

The possible pertinence of an event 250 million years ago to today is twofold: it shows us how triggering events followed by a series of positive amplifiers can issue in devastating results for life, and it calls attention to independently stated concerns today that the contemporary pace of global warming could once again release methane bubbles now covered by ice or sedimented under a cool ocean. Indeed, the temperature increases during that period are close to those now projected for 2100 if nothing radical is done soon to reduce the human input of CO_2 into the atmosphere.[12] One difference, however, is that there are seven billion people on the planet now—up from merely five million in 800 B.C.E. and three billion a mere fifty years ago—and there were none when that great mass extinction occurred. Another even more significant difference is that late-modern capitalism is based upon fossil extractions and CO_2 emissions that trigger large-scale self-amplifiers in the processes of ocean currents, climate, glacier flows, biological evolution, and so on.[13] We are playing with a wildfire, and it is playing with us.

A second event is closer to home in time and mode of life. Around twenty-eight thousand years ago, the last of the Neanderthals were extinguished. If you are a part of a generation over fifty, you are likely to have assumed (as I did) that they disappeared either because they were vanquished by *Homo sapiens* or because they were not intelligent enough to survive the competition for stable supplies of food and resources. We *Homo sapiens* are so smart and violent. That story paints a triumphalist view of humanity (or, better, of one of its types) in a way that recalls old stories of how dumb dinosaurs bit the dust before the dating of that huge asteroid cavity in Mexico. Both stories reenforce a vision of *Homo* triumphalism on the way to new triumphs in the future. They are anchored in the troubled and troubling idea of evolution as ever growing complexity.

The emerging picture is rather different, though it too is replete with uncertainties filled with speculation. According to Clive Finlayson, the *Homo sapiens*-Neanderthal split started six hundred thousand years ago; then each evolved into its "classical shape." They occupied adjacent land for a while in the Mideast, where some intraspecies sex and reproduction occurred. The evolutionary pattern after that date was very bumpy for each type, not corresponding to a smooth image of evolutionary progress.

As the climate cooled 125,000 years ago, the Neanderthal population became concentrated in warmer areas. Why? They had larger brains than *Homo sapiens,* though it may be that the ratio of cerebellum to cerebral hemisphere size differed between the two groups. One Neanderthal site in Jersey (currently an island off the shore of England but once part of an isthmus connecting England to the mainland) reveals that they organized complex hunts for woolly mammoths by forcing them to stampede until they fell over a high cliff; then they cut up, barbecued, and ate them, while warding off other predators. Such a feat involved extensive foresight, improvisation, and organizational skill.[14]

The changing climate and their stocky builds combined to make life tough on Neanderthals. That build had suited them for warm periods when they could stalk and kill large animals on the edge between forests and plains. But cooling wiped out the large animals—or forced them elsewhere—and the Neanderthals were pressed to move south to find food consistent with their body type. They could not, for instance, chase down animals on long hunts or adopt too active a nomadic life.

The last hurrah for the Neanderthals occurred on Gibraltar, an area that retained its warm weather longer. Some had lived there for 125,000 years, as the remarkable layers of temporal evidence on the Gorman Cave show. "By 40 thousand years ago their homeland had been pinned back to the Mediterranean, south-west France, and pockets around the Black Sea. The acceleration of cold and unstable conditions after 37 thousand years ago reduced the range even further, leaving a major stronghold in southern and eastern Iberia and pockets in northern Iberia, the Atlantic seaboard to the north. . . . By the time we reach the end of our period, 30 thousand years ago, the only Neanderthals left were in south-western Iberia."[15]

Between twenty-eight thousand and twenty-four thousand years ago, the remnants became extinct, perhaps from a disease that wiped out the weakened few remaining on the Iberian Peninsula. Finlayson's thesis is that conjunctions between body type, hunting methods compatible with that type, the loss of large animals, and rapid climate change did in the Neanderthals. As he puts it, there was a lot of luck and contingency in how things worked out. By that I think he means that while each of the force fields in play—climate change, animal migration patterns, body type, hunting instruments, type of game available, pace of evolutionary change—was in itself explicable, volatile conjunctions between them created situations that rapidly favored some groups and threatened others. Even if you add, as I would, a teleodynamic element to species evolution that makes each new stage in evolution only partially explicable in advance, the fate of the Neanderthals was shaped by a series of unhappy conjunctions and the variable speeds at which they occurred.[16]

Homo sapiens did not triumph over the Neanderthal in a stable environment, nor did they wipe them out in a series of genocidal battles. Their weak, slim body type, ability to forage and hunt, resultant emergence of village sites of activity, and invention of projectile hunting launched an unplanned growth in the *Homo sapiens* population on the cold plains that helped to protect the species from the ravages of disease and climate change. What if the climate had entered a rapid warming period? Or if that warming had produced new and highly contagious diseases? Or if a new series of volcanoes and methane bursts had occurred? *Homo sapiens* might not have been as lucky as a "species" as it was in fact, though the fate of the Neanderthal might have been severe too.

Are we the Neanderthals of today, packing seven billion people onto the face of the earth during a period when capitalist emissions trigger several amplifiers with lives of their own, climate change is rapid and irreversible, fossil fuel use is still growing, and the prospect of future conflagrations is severe between people trying to emigrate from low-lying zones and militant regimes aggressively protecting somewhat more favorable territories? If the Neanderthals in fact had a narrow line of escape that they did not find or luck into—an idea that is uncertain—then we now have an apparent one that is thwarted by the forces of fossil fuel companies, neoliberal capitalism, the established infrastructure of consumption in old capitalist states, the drudgery of everyday life for so many, evangelical theology in some countries, the limited aspirational projections provided by the media to poor and working-class people, popular denialism, and an abstract belief by many others in climate change disconnected from living drives to political action.

AGGRESSIVE AND PASSIVE NIHILISM

Humanist exceptionalism, to rework an idea drawn from Charles Taylor, is the idea that we are either the one species favored and nourished by a God or an unprotected species so superior to other forces and beings that we can deploy them endlessly for our purposes.[17] We have language, rights, dignity, meaning, creativity, artistry, and endless entitlements; they are either useful to us as "objects" or so inferior that we do not need to think about them intrinsically. The long debate between utilitarianism and Kantianism was largely a debate within humanist exceptionalism. So, too, was that between capitalism and communism. Sociocentrism, cultural internalism, and humanist exceptionalism support one another amid their differences; they provide the shifting matrix within which a series of provincial debates and omissions occur. To varying degrees, all minimize human entanglements with a volatile world that today makes things fragile for the very countries and regions that are most prone to deny entanglement. The contemporary fragility of the old capitalist states can be thought of as eco-blowback, in which the slow violences that extractive cultures have imposed on other areas with desired resources and vulnerable populations now come back to haunt the predatory states, too.

Some did and could argue earlier that humanist exceptionalism formed a bulwark against nihilism. By identifying with humanity, we resist the drive to drain all meaning, purpose, morality, and artistry from life after "the death of God." We invest it in human flourishing. Of course, Nietzsche would not have accepted that reading, because he sees close bonds between monotheism and the humanism it helped to spawn as both opponent and complement. But we do not need to tarry over that well-worn issue.

My question is rather different. Does humanist exceptionalism, in both its divine and secular expressions, run the grave risk of becoming a new mode of nihilism during the era of the Anthropocene? If so, what are the possibilities of overcoming the humanist rush to nihilism?

To pursue this question, I inhabit and revise Nietzsche's account of nihilism. I both draw sustenance from his account and freely work upon it during an era he might have been prescient about but did not anticipate specifically. Nihilism, the sense that all meaning has been subtracted from the world, emerges when a set of urgently needed beliefs and meanings are sorely threatened by events, texts, or experiences. The culturally imbued need might be the belief that God infuses the world with divine meaning, or that Christ will soon return, or that the eternal laws of the world are knowable by Western scientists in principle, or that humanity is uniquely made in the image of God, or that the world is an organic whole to which we belong when we tread lightly on it, or that history is set on a progressive trajectory with possible, positive outcomes we can vaguely discern on a singular horizon, or that humans are the only ones who make a substantive difference on this planet. If and when the culturally infused need is both intense and belief in its actuality is threatened, nihilism emerges as a real force. Nihilism is the sense that one or more of the preceding demands is indispensable to give life meaning but that such a belief is in fact under mortal threat. Because elements of belief and unbelief express differing degrees of complexity within different registers of the body–brain system, and because these subsystems are connected by multiple relays and loops, expressions of nihilism can find different degrees of complexity and self-awareness.

One expression is the emergence of aggressive nihilism, in which the more evidence you encounter that challenges the faith in which you are deeply invested, the more you protect the faith by demeaning, attacking, and punishing those who carry that message. Today the temptation to aggressive nihilism comes not only from threats to the existence of God, or to the loss of organic belonging, or to doubts about capitalist projects of mastery and abundance; it also comes from the palpable sense that the advent of the Anthropocene throws *several* familiar and contending projections of human progress into jeopardy over a short period of time.

Aggressive nihilism responds to shocking evidence that extractive, high-consumption, radically unequal capitalism is self-defeating by radically upping the ante of deniability, by attacking the messengers, and by doubling down on the very activities that exacerbate the problem. "Drill baby drill," the lively, young vice presidential candidate repeated at every election stop in 2008. "Frack, frack, frack us" is the new demand. The most extreme form of aggressive nihilism may be inhabited by the whisper "either we retain what is or we let the whole thing collapse, including perhaps human existence itself." The fossil fuel companies, the right edge of neoliberalism, Fox News, and the extreme wing of evangelicalism take the lead in pushing this agenda in the United States.

The biggest danger of aggressive nihilism is that it will carry the day until it is too late, resisting radical changes in production, consumption, and investment until irreversible changes accumulate too much in glacier movements, ocean acidification, and climate change. But there is a second, more subtle danger attached to it, too. The repeated exposé of aggressive nihilism by its critics, while important, also may lead others to fail to come to grips with a passive nihilism that haunts them. For many of us, too, grieve the loss of a world that can be no longer. I mean to suggest, then, that many still tempted today by the intercoded tendencies to sociocentrism, cultural internalism, humanist exceptionalism, traditional secularism, and anthropocentrism reviewed so briefly earlier are vulnerable to the temptation of passive nihilism during the late period of the Anthropocene.

How does passive nihilism work? The nihilist had internalized a profound need to identify meaning and purpose as such with an engrained set of cultural beliefs and practices. It might be, in one period and place, that the meaning of life as such would become hollowed out if the God who bestows meaning, dignity, morality, promises, and providence were to slip away. Humanist entitlement was an attempt to fill such a need in another way. On a related front, consider how some scientists and humanists responded bleakly after the advent of quantum mechanics. "Nihilism," Nietzsche says, "as a psychological state, is reached . . . when one has posited a totality, a systematization, indeed any organization in all events, and a soul that longs to admire and revere in some supreme form of domination and administration."[18] Your faith in your own value was grounded in a set of conscious and molecular beliefs and now *you* consciously doubt them.

The solid ground begins to rumble. You lose faith. However, the established historical source of meaning had penetrated so deeply that you now assume that *all* meaning and purpose must disappear if this historically salient meaning melts down. We have approached the phenomenon of passive nihilism. Existential anxiety arrives on the way to possible despair.

There is another element, too. "The philosophical nihilist is convinced that all that happens is meaningless and in vain; and that there ought not to be anything meaningless and in vain. But whence this: there ought not to be? From where does one get *this* meaning, *this* standard?"[19] The second-order demand consists of molecular residues persisting within us of the onto-perspective that has been subdued at a more refined cognitive level.

This conflict within and among us becomes a source of anxiety and passivity. How so? As already suggested, both selves and cultures are inhabited by multiple, interacting layers of thought-imbued affect, with cloudy affective purposes and thought-imbued feelings at visceral levels both influencing conduct directly and flowing into more refined layers in ways that strengthen, weaken, or modulate the latter. We communicate directly on the affectively dense molecular tiers, too, through gestures, tonality of voice, the unconscious choice of words, stutters, and extreme responses to minor stress, so that the visceral dimension of life both operates within the social structure of the self and communicates through cultural processes. The multimodal media function as collective amplifiers.

In passive nihilism, second-order residues previously flowing into a set of higher-order beliefs continue to bounce into those refined registers. But now they no longer reenforce and add affective vitality to the latter. Rather, they now create dissonance. Passive nihilists become weakened, anxious beings who doubt a providential God, or the Kantian subject, or human exceptionalism, or sociocentrism, or an organic world, or human mastery over nature. But they/we do not invest in alternative paths of meaning, responsibility, and action because stubborn residues within and between them resist doing so. These other registers, again, consist of culturally imbued proto-thoughts and intensities that communicate with higher zones *and* subsist below the ready reach of argument or simple reflexivity. They subsist as uncanny remainders, like residual attachments to delicious moments with a former lover that may flare up long after you have gotten over the old affair.

A performative contradiction is not in operation here, then, because the variances operate on different registers of subtlety and affective intensity. That is why Nietzsche is so resistant to dialectical reductions of his philosophy. What has happened in this instance is that a stream of connections that formerly was in play across subsystems is now stymied. The cloudy, affectively rich register seeks to connect to the more refined register to which it was wired, but it is now blocked from doing so. Resignation sets in.

There is an old saying in neuroscience which goes "neurons that fire together wire together." A suggestive way to think about the formation of habits, perhaps. The slogan needed here, however, presupposes the first but shows how connections can create dissonance: "subsystems that rumble together sometimes start fumbling together." The flows persist and the dissonance increases. Passive nihilism.

When critics "interpret" these energetic remainders, they are actually offering speculative *dramatizations* that may both touch them and add more form to them than they have. You do not *know* this register in itself, for it lacks the degree of organization needed for knowledge. Just as we initially lack awareness of modes of action on the way even though brain imaging would readily reveal strange agitations to others. Our interpretations, then, dramatize a strain to ascertain whether pretending that the old connections are secure will trouble or activate something in the targeted others. You dramatize to *ascertain which way people turn once this or that dramatization has been offered.*[20] "Tapping" old idols is one way Nietzsche described it.

Because multiple tonalities operate at the molecular level, it may be that positive dramatizations and live role experiments can help to enact new orientations with the greatest promise in a new situation. This, then, is the level Kant anointed in a radically different way as the spontaneous accord of the faculties in the aesthetic judgments that flow forward, that Freud did in another way with his notion of archaic "memory traces" that affect conduct without being known consciously. On the reading advanced here, they are culturally imbued passive syntheses that are not reducible to innate processes, that infuse life and that exceed direct self-knowledge. There is much more to thinking than knowing, though knowing has its importance, too.

Nietzsche tries to touch such uncanny zones through arts of the self; Buddhists through mindfulness and lucid dreaming; Freud through talk therapies; Augustine through confessions in devout tones to clerical authorities. Deleuze collectivizes them through micropolitics; Butler enacts public experiments to attract us to do so, too.

One perplexing issue of passive nihilism during the late stage of the Anthropocene is whether and how to recraft the molecular flows hinged to a past that never was and a future that can no longer be. Because there is never a vacuum on the visceral register of cultural life, and because it packs considerable affective density joined to cognitive crudity, ignoring it is not a good idea. During a major phase transition, we are called upon to consolidate some propensities and dampen others.

My sense is that today the shock of rapid climate change and the massive responses to it required by productive, extractive, consumptive, investment, and exploitative practices of capitalism, its military priorities, its identity demands, and several theological faiths to which it is loosely allied have generated a conundrum. The combination foments seedbeds of passive nihilism in the academy, in consumption priorities of the affluent, in investment portfolios, in some churches, and in political parties. Not denial and aggression but evasion and neglect.

It is a dangerous time. But passive nihilism may tremble with multiple potentialities, some pointing toward an even more morose resignation, others to aggressive nihilism, and others yet to passing through the first two to new investments of meaning, purpose, and drive in a world that no longer fits, if it ever did, the dominant cultural images outlined so briefly earlier.

Nietzsche sensed new possibilities for his era as (he thought) the omnipotent God had been wounded, though he increasingly doubted that many would in fact cultivate existential gratitude for the richness of a world without a providential god. He expected denial and cultural *ressentiment* to carry the day. To take one example of the former orientation, however, in *The Gay Science,* he spoke of the "voluptuousness of a triumphant gratitude that still has to inscribe itself in cosmic letters on the heaven of concepts."[21] Such a voluptuous gratitude rings a bit hollow today, perhaps, and needs reworking. But it does embrace human implication in a world composed of innumerable human and nonhuman forces of multiple types that both enable us and are not intrinsically predisposed to our benefit. He says, in one of his more somber moments, "But to stand in the midst of this *rerum concordia discors* and the whole marvelous uncertainty and rich ambiguity of existence without questioning, without trembling with the craving and rapture of such questioning, without at least hating the person who questions, that is what I find to be contemptible. Some folly keeps persuading me that every human being has this feeling . . . , that is my injustice."[22] Perhaps during an era of climate peril and a future of intense wars and possible extinction, we need to negotiate a more modest set of positive attachments to each other and to the world with which we are entangled through numerous strings, pulleys, and flows, allowing such energies to fuel a new political activism.

Both Pascal and Nietzsche may need to be reworked today. During the Anthropocene, Pascal's wager over the existence of God has morphed into one over how long future projections inside capitalist investments, management practices, resource uses, and class and regional exploitation can be sustained. Nietzsche's voluptuous gratitude for existence in an unruly universe must perhaps morph into new modes of care for the exigencies of an uncertain future during the advent of the sixth extinction.

ENTANGLED HUMANISM

During the Anthropocene, it is certainly not enough to "reject humanism." Indeed, such a rejection, standing alone, is all too apt to add fuel to aggressive nihilism, one that says, "It will be over soon, so let's take care of ourselves as we ridicule modes of political activism anchored in the squishiness of entanglements and care."

Today, perhaps, it is wisest for recovering humanists to transfigure humanist exceptionalism into affirmation of entangled humanism in a fragile world. Such transfigurations acknowledge a world composed of innumerable entanglements, including

- the ancient bacteria imprisoned in our cells (mitochondria); other bacterial *microagencies* in our guts that influence moods flowing into the higher brains
- the olfactory receptors that link our organs together and to the world below sensory awareness and then flow up as microagents into thought-imbued moods and dispositions
- the reptile brains within us that have evolved to forge dissonant lines of communication with more refined brain zones
- the genetic inheritance many carry from the Neanderthal and Denisovan
- myriad modes of interspecies symbiosis and disease jumps
- racial and regional struggles that impinge upon every other entanglement
- planetary, partially self-organizing processes of climate, glaciers, ocean currents, water self-filtration processes, bacterial and viral flows, radiation breakdown, tectonic plates, and species evolution that now impinge upon regions asymmetrically

These entanglements of numerous types are increasingly studied in the ethological sciences. Such knowledges are marked by the mobile modes of remainder, noise, and uncertainty that exceed them. Many of the processes are marked by periodic modes of volatility during their transitions. These, too, infuse and impinge upon entangled humanism.

Some cosmologies of entangled humanism may embrace a god or gods of entanglements; others may be nontheistic; others may pursue an immanent divine; and others yet the theme of an immanent naturalism that is never apt to understand the world completely.[23] Every which way.

The variant of entangled humanism pursued here strives to acknowledge without existential resentment the constitutive imbrications and interdependencies noted earlier: it seeks to negotiate positive affinities of *spirituality* across differences in human *creeds* to both affirm the reality of human entanglement and to act with resolute courage upon the destructive trajectory of hegemonic institutions. The spiritualities of entangled humanism seek to tilt everything they touch by forging a presence on the established field of discourse and practice, without trying to eliminate strains of belief and faith in several other traditions. The pursuit of affinities of spirituality *across* creedal differences is, again, compatible with several versions of theism and nontheism.

Entangled humanists acknowledge limits to the human ability to feel, perceive, think, know, judge, and respond in a world teeming with a variety of human and nonhuman modes of perception that often act upon and exceed us. But, as numerous ethnographies and technoartistic experiments reveal, we do not know in advance exactly where and what those limits are. We may be able to stretch our capacities of receptive experience enough to inhabit the experiential edges of other species, such as whales, vultures, crocodiles, yeast, and crows, without becoming attuned to any completely. Such experiential augmentations may intensify the sense of interspecies entanglements.

To take merely one example, the drive to extinction of the vulture in India is fueled by the addition of a new drug for cattle that is toxic to them. Their extinction will shut down a symbiotic relation whereby humans provide cattle for them to scavenge, the scavenging protects the populace against disease, the cleaned bones provide the poor with items to collect and sell as fertilizer, and the feral dog and rat populations are kept within manageable limits.[24] Is it possible to occupy in our imaginations more of the life world of the vulture? Can, perhaps, experiments in aerodynamic flying, peering through binoculars into a grassland from high perches, tearing meat off a carcass, and the cultivation of new smells be joined to interpretive studies to take us to the edge of vulture experience? I at least need such simulations to overcode an image still circulating in my soul from movies I watched in middle school of "buzzards" gathering silently on a dead tree in the desert, calmly waiting to tear apart a hero or villain on the verge of death.

Entangled humanists are thus wary of transcendental arguments that pretend to fix the boundaries of intra- and interspecies engagement once and for all, given the recurrent defeats of such attempts in the past. As multiply entangled beings, we seek periodically to stretch the visceral habits of perception and identification from which cultural judgments are forged; to work upon congealed drives within and between us by tactical means to open common sense to new modes of experience, distantiation, and attachment; to extend knowledge of other species by confronting new work in allied fields; to enliven the existential dimension of life; and to expose ourselves to new events that may well rebound back on those endeavors.

But what else makes it entangled *humanism*? As we struggle to become worthy of the events we encounter, we also give a recurrently problematized degree of priority to the human species in its interdependencies and imbrications with other beings and forces it neither masters nor owns. We affirm care for the human estate in its worldly entanglements, as that care stretches into regional asymmetries that haunt the world and toward other species with which we are entangled.

Thus the symbiotic relation to the vulture is valued in part because of care for a species that is ugly to many of us and in part because its practices decrease the spread of rabies and anthrax to poor people living in rural India.

Entangled humanists do not seek a morality of purity or a pure horizon of community. Such projections are too closely associated with notions of racism, cultural internalism, nationalism, and extreme human entitlement. To be clear, however, the *humanist* dimension of entangled humanism means that as you become more aware of nonhuman modes of experience and other nonhuman force fields, you also learn that you must cast your lot with some of those processes and species more than others. The difficult question, to be resolved through ethological studies, presumptive care, and experimental action, is, Which ones?[25]

Take a fraught example, there is the salutary recent attempt to save the whooping crane through encouraging chicks to "imprint" upon crane-costumed humans and the use of ultralight aircraft to teach migration patterns to the fledglings. The effort is noble in intention and responsive to the human causes of the potential extinction of that species. But it also poses a series of difficult issues about the use of captive birds for egg laying and the use of other types of cranes as captive egg-sitting surrogates. The point is to dig into interspecies entanglements in each situation to inform experimental action. It is too late simply to leave the world alone.

The questions are, How to renegotiate complex entanglements during the late Anthropocene? And how to construct a militant pluralist assemblage pursuing ecological actions at multiple sites? The latter include revitalizing positive affinities of spirituality across multiple differences in creed; pursuing role experiments to readjust church, consumer, teaching, entrepreneurial, work, scientific, blog, parenting, and electoral priorities to meet the dictates of the Anthropocene; and organizing a cross-country general strike to press states, corporations, churches, universities, localities, consumers, and investors to pursue a series of rapid changes in their institutional priorities.[26] As we—the "we" is pluralized, regionalized, and invitational—move back and forth between new role investments and the other engagements, it is well to remember that the first pursuits make a cumulative difference in themselves when many practice them. *They also work on the molecular dimension of cultural life.* They work upon established modes of feeling, perception, and readiness to act. Micropolitics is thus fundamental to entangled humanism without being sufficient to it. Macropolitics is critical, too; its affirmative possibilities are apt to spring to life when it stays in touch with the positive experimentalisms of micropolitics.

Doubtless several issues remain. I pursue only one now. Do the positive affinities of spirituality across constituencies and regions projected here themselves mix a strain of passive nihilism into the activist mix? Does a subliminal tension between profession and visceral sensibility project the possibility of a new turn in the future when we should in fact come to terms nobly with the irreversibility of human extinction along with several other species with whom that species is entangled?

That is what Guy McPherson, a climate scientist, would say. Extinction of the human estate is already in the cards, he says. The amplifications now under way have become irreversible. It is time to learn how to die with nobility together.[27] McPherson has not yet told one old guy how to inform his children, partner, students, grandchildren, and Facebook friends about that future.

I assume that we are not there . . . yet. Such an assumption is partly grounded in shaky counterevidence, partly in shaky faith in new modes of activism, and, in all likelihood, partly in molecular protests against the prospect of such a future. Perhaps a trace of passive nihilism folds into my perspective. At least I *cling* to entangled humanism in a way that suggests both that the attachment goes well beyond casual belief and that it is threatened during a fragile time. I will do so as long as I can.

Whence, again, comes this last "ought not"? Does it express a clash between previous humanisms still in circulation and a shaky encounter with a new condition, a clash that, if conjoined to care and political experiments of the right sort, could lend support to new, militant pluralist citizen assemblages across regions? I cannot say for sure. Perhaps it is time to give the politics of swarming and cross-regional general strikes a try, focusing demands first and foremost on regimes that have heretofore extracted oil, coal, and lumber from vulnerable areas; produced the most planetary emissions; and imposed intense suffering on marginal zones within and outside those regimes. Then, if that urgently needed improbability proves to be impossible, it may be time to move to the next item on the agenda.

NOTES

1. A preliminary and much shorter version of the first part of this essay appeared as William E. Connolly and Jairus Victor Grove, "Extinction Events and the Human Sciences," *The Contemporary Condition* (blog), July 3, 2014, http://contemporarycondition.blogspot.com/2014/07/extinction-events-and-human-sciences.html. As the original essay was written with Jairus Grove, the chapter in this book is also touched by his thinking. I wish to express my appreciation to Jairus for his work in this general territory.

2. For an excellent review of Marx's flirtation with Epicureanism, and an exploration of the implications of its rejection, see Jane Bennett, *The Enchantment of Modern Life* (Princeton, N.J.: Princeton University Press, 2004).

3. See the last chapter of Weber's *The Protestant Ethic and the Spirit of Capitalism,* trans. Talcott Parsons (1930; repr., New York: Scribner, 1958).

4. For a close, critical engagement with Hayek on the bumpy intersections between capitalist self-organizing processes and those of multiple other processes, human and nonhuman, see William E. Connolly, *The Fragility of Things: Self-Organizing Processes, Neoliberal Fantasies, and Democratic Activism* (Durham, N.C.: Duke University Press, 2013), chapter 2.

5. These tendencies, *by comparison of course to the positions staked out here,* can be found to varying degrees in the following: Hannah Arendt, *Between Past and Future* (New York: Viking Press, 1954), stakes out a position anchored in the idea that the new sciences of quantum mechanics threaten human meaning and enactments. In rough agreement with her, Jurgen Habermas, in *The Future of Human Nature* (Cambridge, Mass.: Polity Press, 2003), argues that the advance of biology and its technologies squeezes out space for human agency. He fails to explore how some complexity theories in biology, by multiplying the zones of microagency within and without, actually deepen our sense of human agency and creativity. On the other hand, a few of those same theories have hitched themselves to a moderate version of neoliberalism. George Kateb, in *Human Dignity* (London: Belknap Press, 2014), limits agency, dignity, rights, and creativity to human beings alone. Walter Benn Michaels, in *Our America: Nativism, Modernism, and Pluralism* (Durham, N.C.: Duke University Press, 1995), attributes racism to those who incorporate a biological element into human culture and philosophies of "difference." He omits reference to thinkers who refuse to identify the biological with fixity or to accept biological reductionism. I offer a critique of that stance in *Neuropolitics: Thinking, Culture, Speed* (Minneapolis: University of Minnesota Press, 2002), though I would now say that the attempt to rethink nature–culture relations there insufficiently addresses a series of nonhuman, self-organizing processes that interact with the human estate. I continue, however, to *think* swarms of difference in ways that transcend the limits he discerns in our ability to *know* difference. He seems, indeed, to restrict thinking to knowing. Ruth Leys, in "The Turn to Affect: A Critique," *Critical Inquiry* 37, no. 3 (2011):

434–72, first seems to separate theories of "affect" from ideational, intentional elements in ways the proponents of the very perspective she criticizes resist. Given this misrepresentation, she can then accuse them of the reductionism and determinism ("anti-intentionalism") they seek to transcend. The result of such misrepresentations is to insulate many cultural theorists from attention to modes of agency within and outside us that help both to constitute and to limit human agency. To help *constitute,* not simply to determine. Several people Leys names as "affect theorists" are in fact proponents of *panexperientialism,* the idea that traces of experience and meaning flow deeply into the biosphere, well beyond the sites of intentional agency acknowledged by advocates of human exceptionalism. Human agency, on this alternative reading, is composed in part by a larger host of microagents below conscious attention: that is one of the things that makes creativity possible. As Leys becomes aware of this position, it will be interesting to see whether she will feel pressed to misrepresent it, too, in order to criticize it. I do not, to date, embrace panexperientialism entirely, but, as the preceding examples make clear, I do concur that multiple modes of meaning, perception, and agency do sink deeply into the biosphere and nonhuman organisms.

6. For a review and critique of the anthropic exception in the natural sciences, see Ilya Prigogine, *Is Future Given?* (London: World Scientific, 2003).

7. Charles Darwin, *The Descent of Man* (1871; repr. New York: Prometheus Books, 1998). For engagements with this dimension of Darwin's work, see Elizabeth Grosz, *Chaos, Territory, Art* (New York: Columbia University Press, 2008), and William E. Connolly, "Species Evolution and Cultural Freedom," *Political Research Quarterly* 67, no. 2 (2014): 441–52.

8. Henri Bergson, *Creative Evolution,* trans. Arthur Mitchell (1907; repr., Mineola, N.Y.: Dover, 1998).

9. My knowledge of Alvarez is mostly derived from Michael J. Benton, *When Life Nearly Died: The Greatest Mass Extinction of All Time* (London: Thames and Hudson, 2005). The next couple of pages bear the imprint of Benton's history of geology as well.

10. See Rob Nixon, *Slow Violence and the Environmentalism of the Poor* (Cambridge, Mass.: Harvard University Press, 2011), and Anna H. Tsing, *Friction: An Ethnography of Global Connection* (Princeton, N.J.: Princeton University Press, 2005). Both of these contributions are discussed in my book *Facing the Planetary: Entangled Humanism and the Politics of Swarming* (2017), of which this essay forms a chapter.

11. Benton, *When Life Nearly Died,* 276–77.

12. See Anthony Barnosky, "A Perfect Storm," chapter 3 in *Dodging Extinction: Power, Food, Money, and the Future of Life on Earth* (Berkeley: University of California Press, 2014).

13. I sometimes adopt the term *Anthropocene* to characterize the recent dynamism between capitalist emissions and the partially self-organizing processes noted earlier. It is not a perfect label, nor are the others with which it is in

competition. In an interesting post titled "The Capitalocene," Jason W. Moore challenges the viability of that name. He dates the take-off period to the sixteenth century rather than to the eighteenth in a convincing way. More fundamentally, he says that those who use the A-term lose explanatory power by talking about industrialism or modernity rather than a more precise source like capitalism. I half agree with him, because those explanatory terms do not do enough work. But those who resist the idea of the Anthropocene miss something, too, because they often seem to assume nonhuman modes of gradualism *until* capitalism entered the scene. To me, the value of the geologic term *Anthropocene* is that, first, it enables us to join geologists and paleontologists who have overturned the thesis of nonhuman *gradualism* that dominated those fields until the 1980s and still serve as background assumptions to much work in the human sciences; second, it identifies periods of relatively rapid change in several nonhuman self-organizing systems even before human beings were important; and third, it thus allows us better to discern how modern capitalist emissions operate as *triggers* to a host of nonhuman *amplifiers* that make the contemporary situation more perilous than the simple idea of nonhuman gradualism later joined to massive capitalist infusions. But what about his charge that theorists of the Anthropocene let capitalism off the hook—hence the call to name it the "Capitalocene"? He makes a good point. But I would refine the terminology even further. Soviet Communism was a major polluter of the atmosphere; Mao pursued the human conquest of nature; classical communist idealism was also one of mastery and abundance; German capitalism has reduced its emissions probably more than any other country; American neoliberal capitalism enters into a resonance machine with evangelicalism that makes it the biggest outlier among the old capitalist states; and on and on. I would complicate the human sources of intervention beyond those advanced by geologists or Moore, using the term *Anthropocene* sometimes and varying it at others. Moore is superb at showing how early capitalism (or precapitalism) started these processes. But instead of looking for "the golden spike," it might be wise to identify several thresholds of inflection, with the greatest acceleration starting in the 1950s. I would also like to see more from him on *how* capitalism intersects with a whole host of self-organizing nonhuman processes. That is, I accept his objections to nature–culture dualism and thus seek examples of the internal dynamisms he supports in principle. Bees are simultaneously capable of impressive self-organization, "serve" as a force of production in human farming, and are at risk of extinction because of farming and other capitalist processes. I don't care which cover term is used, so long as it incorporates within it the multiple dynamisms needed and does not imply that the classical ideal of communism would resolve the issue. It would not. See Jason W. Moore, "The Capitalocene, Part I: On the Nature and Origins of Our Ecological Crisis," *The Journal of Peasant Studies* 44, no. 3 (2017): 594–630, doi:10.1080/03066150.2016.1235036. After this essay was composed, a new book by Jason Moore appeared, *Capitalism in the Web of Life* (New York: Verso, 2015).

Its exploration of how capitalism came to depend on "cheap natures" that are now squeezed is pertinent to issues posed here.

14. See Dimitri Papagianna and Michael Morse, "Meet the Neanderthals," chapter 4 in *The Neanderthals Rediscovered* (London: Thames and Hudson, 2013).

15. Clive Finlayson, *The Humans Who Went Extinct* (Oxford: Oxford University Press, 2009), 127–28. The previous two paragraphs condense several chapters of that book.

16. See Connolly, "Species Evolution," where the contributions of Lynn Margulis and Charles Deacon to a more "teleodynamic" rendering of species evolution are reviewed and clues are drawn from this modified picture of evolution to help us rethink the character of human freedom.

17. This is a key theme in Charles Taylor, *A Secular Age* (Cambridge, Mass.: Harvard University Press, 2007). Taylor construes "exclusive humanism" to be the idea that the virtues and rewards of humanism can be secured without the grace of a loving God. I extend the idea to a mode that can be *either* theistic or nontheistic in shape: the attempt to secure human virtues and capacities without close attention to the efficacy of a variety of nonhuman force fields that have, to varying degrees, dynamic capacities of self-organization that enter into synergies with capitalism and other cultural processes.

18. Friedrich Nietzsche, *The Will to Power,* ed. and trans. Walter Kaufmann (New York: Vintage, 1968), book I, no. 12, 12–13.

19. Ibid., book I, no. 36, 23–24.

20. For a discussion of the indispensability of dramatization to philosophy, specifically as a response to Kantian modes of analysis, see Connolly, *Fragility of Things,* chapter 3.

21. Friedrich Nietzsche, *The Gay Science,* trans. Walter Kaufmann (New York: Vintage, 1974), 2.

22. Ibid., 76.

23. For a salutary and evocative exploration of negative theology in relation to these issues, see Catherine Keller, *Cloud of the Impossible* (New York: Columbia University Press, 2014). Negative theology acknowledges the mystery of divinity all the way down while drawing positive drafts of energy from its impossible pursuit (where the impossible is viewed from merely a cognitive point of view).

24. For a superb exploration of specific species entanglements, including the vulture–human–dog–anthrax–rabies–rat entanglements noted here, see Thom van Dooren, *Flight Ways: Life and Loss at the Edge of Extinction* (New York: Columbia University Press, 2014). Thanks to Jairus Grove for calling this book to my attention.

25. In this respect—and others too—I concur with Thom van Dooren and Donna Haraway. See Haraway, *When Species Meet* (Minneapolis: University of Minnesota Press, 2008).

26. See Connolly, "The Politics of Swarming and the General Strike," chapter 5 in *Facing the Planetary*. It explores movements back and forth between

role experimentations, social movements, and the prospect of a cross-regional general strike. Sorel, Gandhi, Latour, and others are recruited to explore how to transfigure old visions of such a strike—organized around class, building a nation, or opposing a war—to fit cross-regional needs during the Anthropocene. As discussed in that book, a general strike, to work, would have to find expression in numerous regions, including at least India, Bangladesh, Nigeria, Kenya, Antigua, South Africa, Japan and Australia, Europe, and the United States. The vibrant minorities who initiate such actions in different countries will place internal and external pressure upon several regimes at the same time, through refusals of work and reduction of commodity purchases to a subsistence level during the strike period. It will also be a period during which to intensify pressure on a variety of churches, universities, retirement funds, and corporations. This combination is particularly important for the success of such a venture in the United States and Europe, the regions that have to date produced the most carbon emissions and exported many of the worst effects to other areas. It may seem to many as though the regions that suffer the worst effects of those emissions to date are unlikely to join such a movement. That may be true, or it may express the hangover of an old idea that postcolonialism does not need to enter the arena of white environmentalism. Again, Nixon, in *Slow Violence and the Environmentalism of the Poor,* joins others in putting that story to rest. He explores a host of literary activists in Africa, India, Indonesia, and elsewhere who struggle against the slow violence first world capitalism has helped to impose on these regions. He also shows how the pastoral environmentalism spawned in the United States and England does not speak to most ecological circumstances outside those centers, and it is far too narrow an image for them as well. Bhrigu Singh's *Poverty in India: Spiritual and Material Striving in Rural India* (Chicago: University of Chicago Press, 2015) adopts a complementary tack. He places a "minor" tradition in the West into conversation with ecostrategies in rural India, allowing each to become informed by the other. Of course, I know a cross-regional general strike is improbable. The point is that, today, it has become an urgently needed improbability.

27. Dahr Jamail, "Are Humans Going Extinct?," interview with Guy McPherson, *Truthout* (blog), December 1, 2014, http://truth-out.org/news/item/27714 -are-humans-going-extinct?tmpl=component&print=1.

2

Planetary Memories: After Extinction, the Imagined Future

Jussi Parikka

If there is a material, technological, and industrial pollution, which exposes weather to conceivable risks, then there is also a second pollution, invisible, which puts time in danger, a cultural pollution that we have inflicted on long-term thoughts, those guardians of the Earth, of humanity, and of things themselves.

—Michel Serres, *The Natural Contract*

All techniques for reproducing existing worlds and artificially creating new ones are, in a specific sense, time media.

—Siegfried Zielinski, *Deep Time of the Media*

WHEN DOES THE FUTURE BEGIN?

In the *Guardian*-organized live chat in November 2014, the science fiction author William Gibson is asked the rather blunt question by one of the web participants, "When does the future begin?" One could easily become sarcastic at such a broad question, but Gibson refrains from such negativity and takes the question seriously. He observes how the question includes the reference to *future* in lowercase; it comes without the modernist twentieth-century idealization of one big Future waiting for us. Perhaps we are merely in anticipation of lowercase futures, which have lost the vibrancy or energy that was around in the 1980s, Gibson ponders. He continues:

It might represent a kind of very wide cultural maturation. Americans for instance no longer believe in the future as some completely other place. Europeans never believed in that, because in Europe the evidence is all around us that the future is built in the past. We're surrounded by the past in Europe. The American vision of the future was over the hill, down the highway, we'll build a new world. Americans have gotten the message. I think that *Blade Runner* was very important in that, in its wonderfully European depiction of a future Los Angeles that grew perpetually out of its own ruins. A very un-American vision, radically un-American. Something came from that.[1]

Whether it is maturation or just melancholic disappointment remains to be decided. In many ways, the lack of a future can be raised as a dilemma of temporal politics that is haunted by a persistent memory of the past as a sort of a block of imagination; this is what Mark Fisher notes as the *hauntological* tendency of contemporary popular culture and also what is articulated in political philosophies such as Fredric Jameson's. For sure, this cannot be resolved through a nostalgic reiteration or reattachment to a past, but it yet asks for these questions to be asked: What sort of a future and memory of a future are we then able to produce? How do imaginaries of time relate to the theme of extinction as an epistemological and a temporal horizon?

Futures are being imagined constantly, but the emphasis on the spatiality of ruins has been rather central in such narratives of projected futures. A lot of the contemporary imaginary is full of speculations, images, and narratives of Earth before/after humans; the scientific cartographies of the sixth mass extinction are often part of an audiovisual political cartography: the cinema of catastrophes, of the extraplanetary, of futures and future-pasts without humans.[2] Also philosophy and cultural theory engage with the noncorrelated world without us—possibly partly triggered by the certainty of not merely a past before us but also a future without us.[3]

Following Gibson, one can continue to speculate: is it that the other side of this spatialized history—the future as an other place on this planet—is not anymore imaginable in the midst of the encompassing ecological crisis, and as encompassing global financial crisis that encases everything inside its circuit which increasingly are designed as semiautomated or even AI-based computational platforms?

But the question to Gibson had actually one potential meaning that remained unanswered. It was perhaps not so much a question about the future as such as about *when* it might begin. This already places time out of its joint by referring to imagined futures, which turn the historical notion of the trace on its head. If the trace refers to the past, the business of archaeologies of future, to paraphrase Fredric Jameson, is one of "Utopias," their difficult ontological balancing of the existence and nonexistence of the future in the present and the reminder that despite this apparent defiance of logical order, the "not yet being of the future" has to be considered "no less worthy of the archaeologies we are willing to grant to the trace."[4] In other words, any particular future extinction is part and parcel of the current political thought that might be expressed in audiovisual and other media forms, too. One word for that is of course *premediation,* as Richard Grusin has suggested: a form of "medial pre-emption" that works by way of creating a constant atmosphere of possible future scenarios impinging upon us now.[5]

So the future might as well be the now in its uncertain existence, a fact that is underscored by the literal nonexistence of a future for specific forms of life, including humans. I will continue this chapter with two parallel narratives, which structure the argument concerning the possible situations in which the escape velocity of the ecocrisis might unfold as one temporal axis to anchor our discussion in relation to memory, time, and the extinction. Both of the narratives talk of the future and a future-past that is determined in the contemporary scientific and technological imagination. In one way, this imagination could be considered as a sign of the posthistorical—not an epistemological determination that history has ended (as Francis Fukuyama had it) but a recognition of the role of history becoming a programmable object, a mediated narrative, and a media technological context for understanding notions of time that cannot be reduced to the linearly written.[6] This poses an interesting dilemma for extinction that becomes less a predetermined event than a mediated future-present.

The posthistorical is present in different examples of contemporary media and cultural theory. This can be seen as a reference to Vilém Flusser's thoughts in the collection *Post-History.*[7] We will return to those after having presented the parallel narratives that structure for us an idea about different temporalities and what constitutes the present as a form of contemporaneity that sustains the past and future as creative potentialities, not merely the dead rhetorical weight of an inert, spatialized horizon.[8] The concept of posthistory, or "programmed history," is also used in this text to underline the way in which media technological contexts open up to thinking the memory of future-pasts and how this may envelop scientific knowledge production and narrativization in a technological culture facing a cataclysmic collapse owing to ecological crisis. While the intellectual trope of "end of history" has its nineteenth-century precedents in Hegel's philosophy of world history reaching its apex, it is also a mode of thought that pertains to the contemporary situation of geopolitics and the memory of and from the future, after the planetary, after the deluge.[9] In the twenty to thirty years that have passed since the fall of the Soviet Union and the binary world system, the end of history comes to refer less to the "victory" of the liberal order than to the scenario of being unable and unwilling to tackle "the end" of natural history, which becomes domesticated merely as one of the multiple security threats.[10]

But the posthistorical also can be understood through an adaptation of Steven Shaviro's notion of the postcinematic: Shaviro's attempt to understand the aesthetics and politics of audiovisual expression through new forms of cultural dominants can be also tweaked to address the question of the posthistorical.[11] If one cannot claim that history has disappeared, it may no longer be the culturally dominant way of making sense of time or memory. It might, in fact, be in the process of being replaced by modes of thinking that interconnect natural history and "history proper"— connecting the future with the past, the political imaginary with technological fabulation. Besides offering a particular narrative framework, it is also a way to address the variety of temporalities that pertain to a reality conditioned by increasingly sophisticated technologies.

Furthermore, the perspectives presented in this chapter indicate a shift in concepts of time pertinent to making sense of the now. Such a temporal horizon is increasingly discussed in terms of planetary time, including geology and the atmosphere. This realization has found expression in recent cinema culture, through films such as *Into Eternity: A Film for the Future* (2010) and Patricio Guzmán's documentary *Nostalgia de la luz* (Nostalgia for the light, 2010). It has also been the topic of a number of media art projects as well as theoretical discourses: a key example here is the concept of the Anthropocene as discussed in Dipesh Chakrabarty's "The Climate of History," a text addressing the entanglement of global social history and natural history.[12] What all of these projects have in common is a line of argument that reframes the question of the archive and memory in planetary terms. In short, the Anthropocene refers to the discussion in geology if the human involvement in the planet merits a new geological term that follows the Holocene. The discussion has been wide and varied over the past ten to twelve years, but it has already had an impact in the humanities and arts, offering one concept for the sense of agency and its uneven, unequal global nature. This, however, fits in with the ways in which times, past and future, are removed from merely being about the historical time but also the geological time of future. We will return to this idea later also in the context of the notion of the "carbon–combustion complex" that offers a political–economic angle to the issue.

POSTPLANETARY

The first of our narratives of posthistory summons a future. In one of the odd moments offered by Erkki Kurenniemi, the Finnish media art pioneer, he gazes back from the year 2048 without a physical body, without a slimy existence of the flesh. This strange fantasy is itself not without a body, but it recalls the specific historical context of 1980s cyberfantasies: postsingularity, the AI, and the quantum computer—the future is able to reproduce the past as memories for a future mankind that lives in outer space in a digital format that takes a curious place of extinction; extinction becomes actually a threshold in the material form supporting so-called intelligent life in this anthropocentric imaginary.

The brain imagined as software has the temporal span of a different sort of a future than the one limited to our embodied existence. "Software can be pretty much immortal in that good programming solutions and algorithms are really sustainable," to quote Kurenniemi's account from an interview with the film director Mika Taanila. Kurenniemi's vision of cultural heritage is determined by this:

but one clear reason is that we as humans are interested in history. We have museums and we're interested in strange things like archaeology and old music using the original instruments and arranging medieval plays using authentic costumes. We're constantly trying to reawaken the past and IT is a great tool for that, because in fifty or a hundred years when people are interested in the past they will be able to create virtual models of the entire human history. We will be able to transport ourselves into historical reconstructions of different eras in our everyday life. If we'll be able to make the reconstructions work and truly virtual, it will also become an important tool to plan for the future instead of just following some new technology blindly. We can create virtual models of how society will work once it spans the entire solar system and in time, the whole Milky Way. A cloud of golf-ball-sized quantum computer servers, which 10 billion living people could inhabit.[13]

By 2048, Kurenniemi jokes he could be one of those resurrected artificial intelligences looking back. One wonders. What happened? Why did we abandon Earth? Why should the escape velocity discovered in the twentieth century become a vector for a full-fledged narrative of civilization wanting to escape what became perceived as a claustrophobic trap of a planet? The emergence of planetary computation works in parallel to the modern desire or necessity of leaving the planet for other worlds, so often articulated in science fiction in the past but also in more recent productions, such as Christopher Nolan's film *Interstellar* (2014), which is set in an ecocrisis-ridden Earth where the dust storms of the planet trigger a film-length meditation of cosmic dimensions.

Kurenniemi's vision does not seem to give much in terms of (technical) detail, cultural prospects, or political and economic conditions. It is premised on his technical and scientific view of the human being and the brain as a finite automaton that evolution created in its specific slime-based way but that artificial intelligence would show as only one among many possible evolutionary genealogies. Fantasies of reanimation become embedded in the belief of storage capacities and storage power. They resonate with the 1980s visions, but we are constantly reminded that this belief in the technological determination of history has not in any way disappeared. It is an AI-focused way of thinking about time but also a form of reflection that takes into consideration a time of events—a temporal mode that defines future perspective in terms of technological imaginaries where intelligence is deterritorialized from a human capacity to a machinic entity.

The idea is not determined as part of science fiction, but the transformation of intelligence to synthetic intelligence is in operation across the industries of search and networking. In *Wired,* a later contemporary of Kurenniemi, Kevin Kelly, presents a vision of a future Google that is not based on search but on artificial intelligence, enabled by three major technological breakthroughs: (1) cheap parallel computing, where neural network models are seen as neurons of the brain; (2) Big Data and the vast collections of quantified information that constitute an understanding of social life by way of collating massive data in search of patterns that surpass individual volition; and (3) better algorithms to process the data.[14] If one wants to consider Kurenniemi in the context of the contemporary archival mania, one should also expand that investigation into the political economy of the algorithmic AI, because this is becoming yet another way of prescribing the conditions of memory.[15]

However, there is one short text, an interview, in which Kurenniemi pursues the rhetorical trope of leaving the planet. This short meditation complements his long-term vision of 2048, but in ways that offer a political economy of the limited resources of the planetary. In Kurenniemi's "premature self-obituary," "Oh, Human Fart," he discusses the resource basis of a postplanetary future. Kurenniemi's odd relation to environmental thinking produces the idea of turning the planet into "Museum Planet Earth," a fully-fledged planetary preservation program that stops population growth, biosphere changes, and so on. In politically and technologically enforced ways, it sees the end of change, a sort of fabulated end of history, as the solution to the material issues of the planet.[16] The nineteenth-century birth of the museum as a colonial preservation of non-European/Western cultures is extended to the planetary condition.

Kurenniemi's post–welfare state, science fiction economy includes transporting all the other stuff to space: "economic expansion, population explosion, genetic science and nanotechnologies of unimaginable power, warfare." Only a limited number of Earth licenses allow selected people to stay on Earth. Instead, human life as we knew it will be continued in data forms and in space. In a rather fragmented way, Kurenniemi explains the logic of the licenses:

> In 2100, for example, print 10 billion "Earth licences" and distribute them to all the then-living humans. No more licences will ever be printed. Licences can be sold. This way, the people who want long life and long-lived children can have them, but only by migrating into space. This will be cheap, because there will be people wanting to stay down here, purchasing Earth licences at a price that will amply cover the price of the lift to orbit for the seller.[17]

In other words, the mythological desire of leaving the planet—a key feature of Cold War–era science fiction, too—is offset by the ones desiring an unchanging sustainability of the planet, which of course is a parody of the idea of sustainability without change: extinction imagined as a museum condition.[18] Only samples left without reproductive capacities.

A FUTURE *NOMOS*

The second narrative also imagines a future but deals with the geopolitical changes that follow from *staying* on the planet. It is more observant to issues of geopolitics and the asymmetries of having agency in the Anthropocene and is written from a different position as well, despite the somewhat similar future-past perspective. Naomi Oreskes and Erik Conway's *The Collapse of Western Civilization* is a short, fact-based fabulation, a science-fact story of sorts. Subtitled "A View from the Future," the short book written by two historians of science offers an imaginary future written by a future historian: "Living in the Second People's Republic of China, he recounts the events of the Period of the Penumbra (1988–2093) that led to the Great Collapse and Mass Migration (2073–2093)."[19] These events are seen as milestones in a climate change–catalyzed new world order, where the shifting of land and water fronts is a key driver in political changes that Carl Schmitt would have referred to in terms of the *nomos*—the division of the land in political–legal–economic power relations, but which in European legal history was above all a question of troubled relations to the sea, to water.

Since the Renaissance and early modernity, new technologies of measurement from the compass to techniques of mapping were instrumental to the *nomos* of understanding and capturing global space, yet they were always bordering on and negotiating the problem of seas and water, which remained more difficult to measure, map, and divide than land.[20] Hence there is a certain geopolitical irony in that industrially produced global warming causes rising sea levels and the (re)capture of the politically and economically significant dry lands, shifting the *nomos* once more. The once mythical sea now returns in the form of changing legal and governmental borders, architectural plans, and urban planning preparing for a different geophysical future.[21]

The narrator of *The Collapse of Western Civilization* is placed in China, where he observes the chemical features of the industrial revolution. One of the most remarkable aspects of the Anthropocene discussions that have been going on for the past decade has been the recognition that this geological era is also one of massive amounts of chemical dosages. Oreskes and Conway remind us that the planetary placements of CO_2 have been the true hot spots of the past two hundred to three hundred years: the United Kingdom (1750–1850); Germany, the United States, the rest of Europe, and Japan (1850–1980); China, India, and Brazil (1980–2050).[22] The geopolitical order is determined in relation to modes of production, but it also acknowledges the role that the fields of geology and chemistry have played in establishing modern society. This order comes in different temporal shifts, with multiple chemical modernities creating hot spots of production and pollution. Put into the contemporary context, one can add how the differential tempos of the ecological crisis are evidenced in the geopolitical distribution of the waste. This distribution does not necessarily follow the borders of nation-states but becomes clearer in statistics, demonstrating that the majority of emissions come from a restricted number of companies of the "carbon–combustion complex." Among the familiar names of Chevron, Exxon, and BP, one finds the statistic that "the ninety companies on the list of top emitters produced 63 percent of the cumulative global emissions of industrial carbon dioxide and methane between 1751 and 2010, amounting to about 914 gigatonne CO_2 emissions."[23] This refers to the inability to talk of the Anthropocene in the singular as if it were one uniform drive—and not recognizing that it is embedded in the accentuated actions of certain agencies, corporations, and nation-states.

The geopolitical stakes of the planet are readable through the chemical levels, which also affect the heat absorbed in the atmosphere, as we know through various techniques of measurements. The narrative escorts the reader through general facts concerning the political-, scientific-, and policy-related determinations of environmental issues, from calculating the capacity of the planetary sinks—that is, the places where wastes and pollutants end up—to the emergence of practices and idea of "environmentality" or "sustainability." Different political systems respond in different ways, and the narrative reveals the sudden efficiency of the centrally governed Chinese system:

> There were notable exceptions. China, for instance, took steps to control its population and convert its economy to non-carbon-based energy sources. These efforts were little noticed and less emulated in the West, in part because Westerners viewed Chinese population control efforts as immoral, and in part because the country's exceptionally fast economic expansion led to a dramatic increase in greenhouse gas emissions, masking the impact of renewable energy. By 2050, this impact became clear as China's emissions began to fall rapidly. Had other nations followed China's lead, the history recounted here might have been very different.[24]

The planetary temperature rise of up to four degrees had a significant effect in terms of water levels with massive areas of land flooded by the Arctic Sea. Yet the main thrust of the text is not yet another narrative of catastrophic proportions but a meditation on the paradoxical scientific discourse that produced such a situation. Instead of the assumed controversy concerning the interpretation of scientific data, the results concerning causalities of climate change had for years shown a clearly and consistently proven result as to the causes and impact of what was to come. Oreskes and Conway introduce the term *carbon–combustion complex* as a way to make sense of this context: "a network of powerful industries comprising fossil fuel producers, industries that served energy companies (such as drilling and oil field service companies and large construction firms), manufacturers whose products relied on inexpensive energy (especially automobiles and aviation, but also aluminium and other forms of smelting and mineral processing), financial institutions that serviced their capital demands, and advertising, public relations, and marketing firms who promoted their products."[25]

The short book's narrative evaluates the role of public discourse about science in the post–World War II United States and its effect in political decision making in the context of what is labeled market fundamentalism. Since the 1970s and 1980s, neoliberal policies have produced an attitude of skepticism toward scientific positions, which from an economic perspective undermines the specific knowledge perspectives produced by research. This was a radical break with Friedrich Hayek's philosophical neoliberalism, which was founded on a close relationship with the insights provided by research and scientific methods.[26]

The future memory that is being written is at the same time a mix of the most obvious—we know that this is happening, so what's so special about it?—and the most complex: the political, scientific, economic determinations of the geopolitically specific, and yet planetary, dimensions of the sink(ing) ecology. From this perspective, Félix Guattari's concept of "three ecologies"—that there is besides a natural ecology also a social and a mental ecology—sounds almost too innocent as a way of addressing the suicidal, neoliberal capture of the future, even extinction.[27] The collapse of the Arctic ice cap is an ecological event in an ecology of multipliers or active forms that have catalytic effects from chemistry of natural formations of sea, land, and air to economy, urban planning, global politics, security policies, and so forth.[28] The water that was understood as anomalous, or difficult to control and define in the political space of old Europe, becomes once again a determining factor of the geopolitical Earth, but this time because rising ocean surfaces flood coastal areas and metropolises.[29]

Oreskes and Conway's best-seller narrative is parallel to, but also clearly different from, the framing of the planetary in Kurenniemi's visions.[30] Both raise the question of the future memories of the contemporary technological and scientific forces that determine our epistemological and ontological sense of the planetary. However, their differences have to do with accentuated takes on what the planetary as a geophysical entity actually means and how the temporality of the future determines the ecological crisis and extinction as a point of reference that defines the contemporary. Hence I want to turn to a discussion of the contemporary and the posthistorical as significant temporal–political concepts. For it is through these concepts that future-past perspectives gain currency in the evaluation of the political agenda.

In short, Oreskes and Conway's short meditation on the issue of climate change produces an interesting juxtaposition to Kurenniemi's. The future memory produced by the duo and their short novel offers a political economic account of the Anthropocene, even if they don't opt to use this specific term. Kurenniemi's vision is, still, politically undeveloped in contrast to the specific geopolitics that Oreskes and Conway offer—while staying on the planet in contrast to Kurenniemi's postplanetary dreams. The different narratives trigger alternative ways of thinking about the presence of the future in contemporary cultural discussions.

POLITICS OF CHRONOSCAPES

The contemporary moment seems to be increasingly defined by a multiplicity of times and the various ways in which we are trying to make sense of these multitemporalities, or chronoscapes, to use Sarah Sharma's term.[31] It is against the backdrop of such a chronoscape that the entities of a "politics of nature"—most notably the various expressions of climate change, from global warming to changing chemical balances in air, soil, and oceans to the threat of mass extinction—are to be judged. The key premise of this chronoscape is, as already noted, that ecocrisis is not just a present dilemma but a future that acts on the now.

In terms of the notion of the contemporary, the narratives presented herein are ways to get us thinking about the multitemporal stakes of this political category, so significant for modern politics.[32] They involve implicit and explicit ways to deal with ideas of programmed futures, future-pasts, and the agenda of posthistory, which has penetrated the political scene since the 1990s at least. In the postcommunist era after the fall of the Berlin wall, the Soviet Union, and other institutions and symbols of the Cold War era, discourses of end of history also emerged.[33]

This popular—and neoconservative populist—sense of temporality paralleled the rise of various projects, discourses, and corporations of global digital culture. Kurenniemi's ideas were partly a product of the same historical period, whereas the more recent, ecological narrativizations are the next phase of an approach that may be called "posthistorical": it ranges from popular culture examples such as the documentary series *Life after People* (2008) to the scientific discussions of the Anthropocene and such critical insights in fiction and scholarly work like Oreskes and Conway's. In some popular cultural narratives, such as the film *Interstellar* (2014), commentators such as George Monbiot perceive a melancholia of political helplessness that he labels a "politically defeatist fantasy of leaving the planet."[34] One could easily see this relating to key features of Kurenniemi's thought and as part of a longer history of science fiction of underground and extraplanetary life.[35] However, to be clear, *Interstellar*'s view of temporality of the planetary condition is not restricted to a future perspective of leaving (the future as an alternative place to be occupied): it is a twist on the familiar Spielbergian meditation on the crisis of the family system, seen in terms of the cosmic dimensions of the ecocatastrophy and time-critical relativity theories.

But a key argument of this chapter is that the concept of the posthistorical refracts into multiple historical and temporal ecologies that are not merely linear directions but atmospheres of time. At this juncture, the discussion of time and its involvement in the planetary political crisis is one of the most important theoretical issues to consider. One would imagine that recent debates on accelerationism could work in this direction, for at some implicit level, the 1990s cyberfantasies of Nick Land respond to the future-oriented singularities of Kurenniemi. The difference is that Land produces a more explicit thematization of the "forward investment in the future" and the cybernetic mutation of the body.[36] The posthistorical comes out also in the versions of accelerationism that try to execute a determination of the contemporary moment through fabulations about a capitalist future of nonhuman, cybernetic, artificial intelligence. These latter, less delirious fantasies are premised on a temporal scheme that thinks in terms of future-pasts, while taking into account climate crisis–ridden and economically stagnating capitalist contexts, as well as the crises that ensued post-9/11 and the series of economic crashes and austerity measures marking the last decade.[37]

But this is not the only sort of temporal determination that is able to engage with a governmentality of the planetary or with a politics of time and the political imaginary of a future memory. The current discussions concerning the Anthropocene, or the microtemporalities of media culture, refer back to an idea of the variety of temporalities that are constantly synchronized in relation to a horizon of what we could call the contemporary and that might inform our way of understanding the present. It's in relation to this body of theory that Wendy Brown articulates her concise theory of the highly significant temporal determinations of the political.[38] Notions of genealogy, hauntology, and other temporal concepts emerging in works of cultural theory, from Freud to Benjamin, Foucault, and Derrida, are indispensable for the political vocabulary of modernity.

The importance of the genealogical has been already incorporated in much of contemporary media theory—especially media archaeology—in ways that resonate with Brown's articulation of the task of the genealogical method: "to denaturalize existing forces and formations more thoroughly than either conventional history or metaphysical criticism can do."[39] But if the genealogical method opens up the past in terms of "faults, fractures, and fissures," as critical media histories have done to demonstrate the scientific and technological determinations of the now, might there be a way to expand this focus to take into account the multitemporality of our contemporary moment?[40] Such a possibility is already implied in the genealogical method, in the sense that it is a "political ontology of the present" (as Brown states in reference to Foucault).[41] But the contemporary can be seen as a further elaboration of the immanence of temporality to a material context as well as the "questions, meaning, or projects" that invest it.[42] Brown draws on Walter Benjamin's theses on history as a way to develop a political notion of time that is all at once a critique of notions of linear progress, Rankean objectivity (approaching history "the way it really was"), and other reductionist approaches to the temporality of the contemporary. But implicitly, it also raises the question of how to further develop a political theory grounded in complexities of time, with respect to a situation when our relation to the future—even the anthropocentric imaginary of no-future—is also prescribed by science, technology, and media culture.[43]

With reference to Brown's theoretical elaborations and Sharma's ethnographic research, I want to underline the possibility of thinking about the contemporaneity of the present as informed by multiple temporalities and synchronization across the time scales. The rethinking of social temporalities and memory proceeds by way of an entanglement of narratives, material contexts, and a recognition of the different ways in which the future imagined becomes a questioning of what the present contemporary actually is. Sharma's emphasis on power chronographies becomes a way of accounting for the differentially existing timescapes that are produced in relations of labor, gender, and ethnicity and, broadly speaking, in the geopolitics of contemporary capitalism. The critics that argue that homogenization of time is one of the characteristics of capitalism miss out on this more nuanced perspective on capitalism's multitemporal operational logic.

Sharma's ethnographic methodology offers ideas to a wider cultural analysis of time, media, and capitalism. It also brings a different angle to discussions of social memory. In many ways, the contemporary context for imagining future memory has been influenced heavily by the presence of a variety of concepts of *longue durée* that prescribe futures of apocalyptic proportions. The environmental crisis in particular unfolds as a production of discourses of sustainability and apocalypse, yet both are unfulfilling when it comes to handling the complexity of the situation. Rhetoric of sustainability, which dominates current policy making, is not able to question the more fundamental political and economic stakes in the situation. An apocalyptic rhetoric is, for its part, in danger of undermining all sense of agency, producing melancholic forms of subjectivity deprived of capacity for action.[44]

It's clear that we need more effective ways of making sense of the contemporary, drawing on an imaginary future and its pasts. A more satisfying solution is to think of the uneven and multiple overlapping temporalities that help to determine the otherwise broad concepts of the *political contemporary*. Indeed, in the context of discussions of the planetary, the Anthropocene, one is constantly reminded that the narratives of the contemporary technological condition produce multiple temporalities as their ecology of time. It is clear that Kurenniemi's type of narrative differs from the more ecologically minded narrative of Oreskes and Conway, despite the superficial parallels. Indeed, the concept of a sensitive coexistence of many times is a way of approaching a political imaginary of time where the projections of the future that derive from computer simulations of climate crisis and its effects (let's say, the changing temperature of the planet) are *already* acting on levels that entail different temporalities: the time-critical operations of computerized epistemologies, the narrative prescriptions of possible futures, the political decisions based on such data, and so on. Instead of the cybercritique of homogeneous cybertime or the homogenization of time in policy, one should actually emphasize the multiplicities of time as a way to grasp the link between the planetary and the computational.[45]

Indeed, one can reveal a range of micro- and macrotemporalities that govern the future-past temporalities of the posthistorical. Any determination of the "post" of history has in this sense to become true to the understanding of technologies and techniques of time relevant to our sense of historicity. The posthistorical reveals itself through other instances than the historical writing and production of time. Hayden White's concept of "metahistory" was important for understanding writing as a media technology that was as essential to the historical epistemology informing modernity. But it is equally important to understand Wolfgang Ernst's media archaeological emphasis on the microtemporal dimension of machinic time: the various concepts of time that result from a close analysis of the circuits of cybernetic machines show us that there is a fundamental difference between the older techniques of keeping time (calendars, watches, and so on) and machines that automatically produce their own timings.[46]

Vilém Flusser's idea of posthistory might then be the necessary link between the various approaches to the future past, even if it entails taking Flusser beyond the original framework of his thinking. Now, the idea of the programmed dimension of posthistory is not envisaged as a postmodern collage but is identified in the various applications and platforms of computation, in which time is bent and twisted in a variety of ways that resurface as distinct alternatives to history-writing.[47] The posthistorical is a concept of time and politics that arises once we pay attention to the actual functions of a technical apparatus removed from a programmer's intentions, argues Flusser. We can develop this claim so that its concept of "posthistory" becomes a key epistemological framework for the future-past as well. Flusser notably reminds us that to understand the programmability of time–history–memory, "we must neither anthropomorphize nor objectify the apparatus."[48] In other words, approaching the issue of the future-past and the geopolitics of capitalism does not necessitate a perspective of monorail temporality but carefully analyzes multiple temporalities that in technical and epistemological ways narrate the future as an archaeological existence of projected spaces of potentiality.[49]

CONCLUSIONS

In Maurice Halbwachs's classic accounts of memory, he reminds us how memory always takes place in and across collectives.[50] Memory is never determined only as an individual affair, but it always takes place among strangers: the collective practices, techniques, and technologies of passing on cultural repetition are a way of sustaining a sense of the collective—memory and its collective basis are coindividuated. It is, however, extremely important to underline that the list of strangers making up memory is longer than we may imagine: with new forms of communication media, it becomes extended to new platforms, techniques, and habits. The strangers who are our memory and help to propagate it exist in the middle of a circulation of information, goods, and people, of governmentalities far beyond those of nation-states or other institutions of planetary significance (whether security and intelligence agencies, NASA, or, for example, some standard bodies of global governance). This memory, however, is also projected as a collective imaginary of future(s)-present.

Discussing any contemporary analysis of techniques of memory—whether its platforms, practices, or technologies—one is forced to ask how this contemporaneity produces its own pasts, presents, and futures. In this chapter, I have tried to address this issue through the alternative narratives of a future-present engaging the contemporary moment of ecocrisis and technopolitics, extinction, and planetary scales. Those narratives compel us to consider the cultural politics of time as one of geopolitics and a multiplicity of times, from the imaginary of technological futures of the outer planetary (Kurenniemi) to the ones tightly narrated as part of the changing *nomos* of the planetary and climate change (Oreskes and Conway).

Indeed, in the sense that temporal concepts like the genealogical became important for a politics of and out of history (to use Brown's phrasing), we are facing a crucial ecological task of creating vocabularies of the future to make sense of the contemporary post-9/11, post-2008-bank-crash, post-the-contemporary-catastrophic-ecological-crisis, and post-capitalism that sometimes defies notions of history and insists on the necessity to go back to talking in terms of natural history: of geological periods and durations without humans.[51] This is not meant to naturalize the contemporary cultural or economic situations but to demonstrate how a cultural politics of time also includes relations to the nonhuman. To return to the earlier point, cultural heritage, cultural memory, and social memory are increasingly debated in relation to the planetary, the geological, and the Anthropocene; these involve various chemical, geological, and biological processes that move further from the usual parlance related to the social and remind us of the various ecological materialities that determine the times we are living in and toward.

NOTES

1. William Gibson, "William Gibson Webchat," *Guardian,* November 24, 2014, http://www.theguardian.com/books/live/2014/nov/21/william-gibson -webchat-post-your-questions-now.

2. Already in the 1980s, Giuliana Bruno's early reading of postmodern culture and Los Angeles/*Blade Runner* spoke of the "the postindustrial city [as] a city in ruins," characterized by a loss of history in the modern sense of the trope that gives a structured sense of place, agency, and meaning. Imagined futures were starting to be embedded in a state of melancholy of the imaginary surrounded by a sense of posthistorical loss of grand stories. Bruno, "Ramble City," *October* 41 (Summer 1987): 65.

3. Quentin Meillassoux's key work, *After Finitude,* offers the philosophical idea of the *arche-fossil* that signals a world before humans. Meillassoux, *After Finitude: An Essay on the Necessity of Contingency,* trans. Ray Brassier (New York: Continuum, 2008). Besides such a temporal figure at the center of the contemporary philosophy discussions, one finds a wider set of arguments for nonhuman realities in speculative realism. In parallel to such temporal figures as Meillassoux's, Ray Brassier speaks of the "truth of extinction" that triggers the necessity to address "time altogether without thought." See Steven Shaviro, *The Universe of Things: On Speculative Realism* (Minneapolis: University of Minnesota Press, 2014), 74. In addition, Eugene Thacker summons in his philosophical take the occult quality of reality as one that "is indifferent to human knowledge." Thacker, *In the Dust of This Planet: Horror of Philosophy, Vol. 1* (Winchester, U.K.: Zero Books, 2011), 53–54.

4. Fredric Jameson, *Archaeologies of the Future: The Desire Called Utopia and Other Science Fictions* (London: Verso, 2005), note 12, xv–xvi.

5. Richard Grusin, *Premediation. Affect and Mediality after 9/11* (Basingstoke, U.K.: Palgrave Macmillan, 2010).

6. See Francis Fukuyama, *The End of History and the Last Man* (New York: Free Press, 1992).

7. Vilém Flusser, *Post-History,* trans. Rodrigo Maltez Novaes (Minneapolis, Minn.: Univocal, 2013). Flusser's notion relates to civilizational thresholds, referring to the ontological regimes of agrarian and industrial society with their specific relations to time.

8. Wendy Brown, *Politics out of History* (Princeton, N.J.: Princeton University Press, 2001), 171.

9. See Fredric Jameson, "The End of Temporality," *Critical Inquiry* 29, no. 4 (2003): 695–718.

10. See Fukuyama, *End of History.*

11. Shaviro, *Post-Cinematic Affect* (Basingstoke, U.K.: Zero Books, 2010).

12. Dipesh Chakrabarty, "The Climate of History: Four Theses," *Critical*

Inquiry 35 (Winter 2009): 197–222. See also Jussi Parikka, *A Geology of Media* (Minneapolis: University of Minnesota Press, 2015).

13. Mika Taanila, "Drifting Golf Balls in Monasteries: A Conversation with Erkki Kurenniemi," in *Writing and Unwriting (Media) Art History: Erkki Kurenniemi in 2048,* ed. Joasia Krysa and Jussi Parikka (Cambridge, Mass.: MIT Press, 2015), 298–99.

14. Kelly's ideas about the emerging AI world do not, however, make the same rhetorical mistake as Kurenniemi; he emphasizes that these are not dreams of a singularity but of more enhanced smart services that proceed by way of algorithmic reasoning and massive investments, quoting these figures: "According to quantitative analysis firm Quid, AI has attracted more than $17 billion in investments since 2009. Last year alone more than $2 billion was invested in 322 companies with AI-like technology. Facebook and Google have recruited researchers to join their in-house AI research teams. Yahoo, Intel, Dropbox, LinkedIn, Pinterest, and Twitter have all purchased AI companies since last year. Private investment in the AI sector has been expanding 62 percent a year on average for the past four years, a rate that is expected to continue." Kevin Kelly, "The Three Breakthroughs That Have Finally Unleashed AI on the World," *Wired,* October 27, 2014, http://www.wired.com/2014/10/future-of-artificial-intelligence/.

15. On Kurenniemi and social media culture, see Eivind Røssaak, "Capturing Life: Biopolitics, Social Media, and Romantic Irony," in *Writing and Unwriting (Media) Art History,* 213–24.

16. Erkki Kurenniemi, "Oh, Human Fart" in Krysa and Parikka, *Writing and Unwriting (Media) Art History,* 5–9.

17. Ibid, 9.

18. John Beck and Mark Dorrian, "Postcatastrophic Utopias," *Cultural Politics* 10, no. 2 (2014): 132–50.

19. Naomi Oreskes and Erik M. Conway, *The Collapse of Western Civilization: A View from the Future* (New York: Columbia University Press, 2014), x.

20. Bernhard Siegert, *Passage des Digitalen: Zeichenpraktiken der Neuzeitlichen Wissenschaften 1500–1900* (Berlin: Brinkmann and Bose, 2003), 65–120.

21. "Law precedes science and perhaps engenders it; or rather, a common origin, abstract and sacred, joins them. Beforehand, only the deluge is imaginable, the great primal or recursive rising of waters, the chaos that mixes the things of the world—causes, forms, attributions—and that confounds subjects." Michel Serres, *The Natural Contract,* trans. Elizabeth MacArthur and William Paulson (Ann Arbor: University of Michigan Press, 1995), 53.

22. Oreskes and Conway, *Collapse of Western Civilization,* 2.

23. Suzanne Goldenberg, "Just 90 Companies Caused Two-Thirds of Man-Made Global Warming Emissions," *Guardian,* November 20, 2013, https://www.theguardian.com/environment/2013/nov/20/90-companies-man-made-global-warming-emissions-climate-change. Goldenberg's article is referring to Richard Heede, "Tracing Anthropogenic Carbon Dioxide and Methane Emissions to Fos-

sil Fuel and Cement Producers, 1854–2010," *Climatic Change* 122 (2014): 229–41.

24. Oreskes and Conway, *Collapse of Western Civilization,* 6.

25. Ibid., 36–37.

26. Ibid., 43.

27. Félix Guattari, *The Three Ecologies,* trans. Ian Pindar and Paul Sutton (London: Athlone Press, 2000).

28. Keller Easterling, *Extrastatecraft: The Power of Infrastructure Space* (London: Verso, 2014), 95.

29. Carl Schmitt, *The* Nomos *of the Earth,* trans. G. L. Ulmen (New York: Telos, 2006).

30. Kurenniemi's world is closer to the familiar discourses of singularity in science fiction: the idea that technological progress will produce a threshold moment when artificial intelligence will rapidly emerge as a significant new sort of world-changing entity that has major impact in terms of the human world. Writers interested in singularity include Ray Kurzweil and Vernor Vinge, and also Charles Stross; the subject has been discussed since the 1980s. Having said that, an earlier context for the term emerges in the work of John von Neumann and his concern for the singularity, as quoted by Stanisław Ulam: "One conversation centered on the ever accelerating progress of technology and changes in the mode of human life, which gives the appearance of approaching some essential singularity in the history of the race beyond which human affairs, as we know them, could not continue." Ulam, "John von Neumann, 1903–1957," *Bulletin of the American Mathematical Society* 64, no. 3 (1958): 5. It's this earlier computer science context that was Kurenniemi's reference point, too.

31. Sarah Sharma, *In the Meantime: Temporality and Cultural Politics* (Durham, N.C.: Duke University Press, 2014). Also the notion of "contemporary" in contemporary art discourse triggers the presence of multiple temporalities; it acts as a marker of time that distinguishes contemporary art from modern but also has implicitly inside it the way in which it consists of multiple temporalities, as Peter Osborne demonstrates in *Anywhere or Not at All: Philosophy of Contemporary Art* (London: Verso, 2013). In other words, there would be a bigger parallel discussion between the temporalities in contemporary art projects and what I present here, but it has to wait for another context to be addressed.

32. Kia Lindroos and Kari Palonen, eds., *Politiikan Aikakirja. Ajan Politiikan ja Politiikan ajan Teoretisointia* [The time book of politics: Theorizing time, politics, and the politics of time] (Tampere, Finland: Vastapaino, 2000).

33. See Fukuyama, *End of History.*

34. George Monbiot, "*Interstellar*: Magnificent Film, Insane Fantasy," *Guardian,* November 11, 2014, http://www.theguardian.com/commentisfree/2014/nov/11/interstellar-insane-fantasy-abandoning-earth-political-defeatism. Monbiot's notes resonate on some level with the political critique summoned by Jameson: "Confusion about the future of capitalism—compounded by a confidence in technological progress beclouded by intermittent certainties of catastrophe and

disaster—is at least as old as the late nineteenth century, but few periods have proved as incapable of framing immediate alternatives for themselves, let alone of imagining those great utopias that have occasionally broken on the status quo like a sunburst." Jameson, "End of Temporality," 704.

35. See Beck and Dorrian, "Postcatastrophic Utopias."

36. Robin Mackay and Armen Avanessian, eds., introduction to *#Accelerate: The Accelerationist Reader* (Falmouth, U.K.: Urbanomic, 2014), 42.

37. Ibid., 43; Alex Williams and Nick Srnicek, "#Accelerate: Manifesto for an Accelerationist Politics," ibid., 347–62.

38. Brown, *Politics out of History.*

39. Regarding media archeology, see Thomas Elsaesser, "The New Film History as Media Archaeology," *CINéMAS* 14, no. 2–3 (2004): 71–117, as well as Jussi Parikka, *What Is Media Archaeology?* (Cambridge: Polity, 2012). Brown, *Politics out of History,* 103.

40. Brown, *Politics out of History,* 103.

41. Ibid., 104.

42. Ibid., 161.

43. Thus it is no wonder that recent political and cultural theory has increasingly turned to acknowledging such aspects of the future as significant for a post-9/11 world of media-informed cultural politics: I am here thinking of Brian Massumi's work on the future-anterior; Richard Grusin's concept of premediation; and, for example, Greg Elmer and Andy Opel's work on preemptive security strategies. Albeit with different emphases, all work upon the same terrain of the future that is constantly present, whether as an atmosphere of fear (Massumi) or as constantly premediated and prescribed, and through such narrative techniques of controlled potentiality (Grusin). Massumi, "The Future Birth of the Affective Fact: The Political Ontology of Threat," in *The Affect Theory Reader,* ed. Melissa Gregg and Gregory J. Seigworth, 52–70 (Durham, N.C.: Duke University Press, 2010); Elmer and Opel, *Preempting Dissent: The Politics of an Inevitable Future* (Winnipeg, Canada: ARP Books, 2008); Grusin, *Premediation.*

44. Indeed, this risk pertains to at least some forms of accelerationism, especially Nick Land's odd version of Deleuze and Guattari, which offers a version of world history determined from the future perspective of the AI Capitalist World Order, or the dissolved human cultures that are emerging in the forces of inhuman cognition and technosentience, to use Land's terminology. Land, "Circuitries," in Mackay and Avanessian, *#Accelerate,* 255. Land's ideas seem to resonate with Kurenniemi through the rhetorical gestures acknowledging the deterritorizaliation from the slimy human body cognition to a technics thinking itself. Despite the future-past of this vision and quasi-radical rhetorics, it remains short of offering a complex notion of time that would account for the uneven and constantly contested distribution of time and planetary resources alongside exhaust. It becomes a monorail approach to distribution of time and other planetary resources, without acknowledging the differential status of how the contemporary is being allocated. See also Doreen Massey, *For Space* (Thousand Oaks, Calif.: Sage, 2005).

45. Wendy Chun speaks of the (computer-)modeled aspect of time in terms of the software ontology of our programmed knowledge of the future. This is most clearly stated in her analysis of the simulations concerning global temperatures and carbon emissions where projections build on existing historical data. In her words, "the weirdest and most important thing about their temporality is their hopefully effective deferral of the future: these predictive models are produced so that, if they are persuasive and thus convince us to cut back on our carbon emissions, what they predict will not come about." Wendy Hui Kyong Chun, "Crisis, Crisis, Crisis, or Sovereignty and Networks," *Theory, Culture, and Society* 28, no. 6 (2011): 107.

46. Wolfgang Ernst, *Digital Memory and the Archive,* ed. Jussi Parikka (Minneapolis: University of Minnesota Press, 2013), 30.

47. Ibid.

48. Ibid., 26.

49. On narrating as counting, see Ernst, *Digital Memory and the Archive,* chapter 1.

50. Maurice Halbwachs, *On Collective Memory,* ed. Lewis A. Coser (Chicago: University of Chicago Press, 1992).

51. See Tiziana Terranova, "The Red Stack Attack," in Mackay and Avanessian, *#Accelerate,* 379–99.

Figure 3.1. Joanna Zylinska, from her film *Exit Man,* 2017.

3

Photography after Extinction

Joanna Zylinska

This chapter takes the horizon of extinction as a reference point against which I will think the ontology of photography and its agency. In the argument that follows, I will explore what photography can do *with* and *to* the world, what it can cast light on, and what the role of light is in approaching questions of life and death on a planetary scale. Arising out of the geopolitical sensibility encapsulated by the term *Anthropocene,* a geological epoch in which the human is said to have become an agent whose actions have irreversible consequences for the whole planet, the "after extinction" designation of my title is not aimed at envisaging a future time when various species, including the one we are narcissistically most invested in—ourselves—have disappeared. Rather, it points to the *present* moment, a time when extinction has entered the conceptual, visual, and experiential horizon of the majority of global citizens, in one way or another. Extinction will thus be positioned as a looming affective fact: something to be sensed and imagined *here and now.*[1]

Thinking photography under the horizon of extinction will allow me to draw two temporal lines in the history of this particular medium: one extended toward the past, the other toward the future. Considering the history of photography as part of the broader nature–cultural history of our planet, I will trace parallels between photographs and fossils and read photography as a light-induced process of fossilization occurring across different media. Seen from this perspective, photography will be presented as containing an actual material record of life rather than just its memory trace. But I will also go back to photography's original embracing of the natural light emanating from the sun to explore the extent to which photographic practice can tell us something about energy sources and about our relation to the star that nourishes our planet. I aim to do this via an engagement with photographers who have consciously adopted the horizon of extinction as their workspace—from the nineteenth-century geologist-photographer William Jerome Harrison through to contemporary artists such as Hiroshi Sugimoto and Alexa Horochowski. I will also look at the postdigital practice of Penelope Umbrico, in which the work of the sun has been taken up as both a topic and a medium.

Yet, in a certain sense, we have *always* lived in the time "after extinction" and hence under its horizon. Five mass extinctions are said to have taken place during the history of our planet, each wiping out significant populations of living beings. At the same time, extinction is first and foremost a process rather than an event: it is an inextricable part of the natural selection that drives evolution. Indeed, geologists talk about "background extinction," a prolonged course of action unfolding across scales of cosmic time, with the average "background extinction rate" of mammals roughly calculated as 0.25 per million species-years.[2] Mass extinctions—of which we are currently said to be awaiting the sixth—can therefore be described as moments of intensification on the timeline of continuous expiratory duration. But even though we have always lived "after extinction," extinction did not enter our conceptual spectrum until the eighteenth century, when it was brought in to provide an explanation for the existence of fossils for which no living correlates could be found. All the scientific explanations notwithstanding, the awareness of extinction as a biogeological fact still does not seem to have become fully embraced by the human population. Biologist Ilkka Hanski claims that, because of our "cognitive incapacity to perceive large-scale and long-term changes," our present grasping of significant geological transformations is very limited. And, as "the apparent stability of the current state of the world is deceiving our senses," we have failed to develop a responsible long-term response to climate change.[3]

This is why the looming prospect of the human-influenced sixth extinction is something we can so easily remove from our consciousness, even if we have heard the facts, seen the simulations, and studied the data. Dwelling under the horizon of extinction without turning our gaze away from it therefore presents itself as an ethical task, I suggest—as well as a condition of any meaningful nonparochial politics. But if we are to devise a truly cosmic political project—for the humans of here and now but also for our human and nonhuman descendants—we need to force ourselves to combat our cognitive and sensory limitations to grasp extinction not just as a concept but also as a set of material conditions. The exercise in imagining extinction, including the extinction of our own species, could be a first step in this process. Naturally, there is something self-defeating about this exercise in philosophizing across cosmic scales: it forces us to acknowledge that, after the extinction of the human, once the rats or the microbes have inherited the Earth, there will be no one to do philosophy or art—at least in the way we humans have shaped these practices. This is not to suggest that the event of human extinction will be more dramatic for the ecosphere as a whole than the previous extinctions (dinosaurs will of course have "cared" about the Cretaceous–Tertiary mass extinction much more) but rather to reintroduce the human standpoint to the theorization of extinction as a concept and a problem. I am proposing to do this not to inject a dose of humanism into our debate but to avoid what Donna Haraway has criticized as a "view from nowhere"—a view that ends up smuggling back the (usually white, straight, male) human into the debate under the umbrella of its supposed nonhuman perspective.[4] With this reservation in mind, I would like to pick up the exhortation issued by the Center for 21st Century Studies at the University of Wisconsin–Milwaukee to "think of the event of extinction not as destructive or final, but as generative," and to consider what happens to writing, theory, and philosophy—but also to art and photography—as devices for enabling a radically different set of arrangements for the world (aka a radically different politics) *after thinking the event of extinction.*[5] Jan Zalasiewicz, a U.K. expert on the Anthropocene, claims that the extraordinary geological significance of our current period in which the human has become an agent of geological change will be reflected in the fossils left to future generations.[6] As Elizabeth Kolbert aptly highlights with this rather humbling image from Zalasiewicz, "a hundred million years from now, all that we consider to be the great works of man—the sculptures and the libraries, the monuments and the museums, the cities and the factories—will be compressed into a layer of sediment not much thicker than a cigarette paper."[7]

It is the question of the material manifestation of this geological significance of our current period that is of particular interest to me here, precisely because it allows us to confront the transience of our human needs, desires, and memories with the more permanent record of human and nonhuman life that endures in time. Indeed, if we think in terms of deep history, we can say that the past leaves an imprint of itself in the rocks, or even—although this may perhaps yet still seem like an association too far—that the past photographs itself. Yet, in what follows, I will argue that the link between fossilization and photography is more than just a metaphor and that this conceptualization can tell us something new both about the photographic medium and about its conditions, *which are also the conditions of our existence: light, energy, the sun.*

The manner of thinking about media in geological terms inscribes itself in what Jussi Parikka identifies as the wider "drive toward geological and geophysical metaphors in media arts and technological discussions."[8] This drive can be accounted for by the fact that science itself "implicitly perceived the earth as media," analyzing as it did, and still does, fossils in terms of "records," "indices," and "biofilms."[9] Writing, reading, and interpretation therefore seem to reside at the very heart of what have become known as Earth sciences.[10] Indeed, John Durham Peters explains that for both Darwin and his close friend Charles Lyell—known as the founder of geology—"the earth is a recording medium."[11] Yet Peters also reminds us that "knowledge is necessarily historical, even in sciences where history might seem irrelevant. The universe is a text, a distorted text, that comes from afar—a classic hermeneutical situation."[12] This conclusion should not be misread as an attempt to "reduce" everything to textuality but rather as an intimation that the universe presents itself to us through the tropes, tools, and media we ourselves have forged as part of our becoming-human. (It *may* indeed present and reveal itself entirely differently to different species or classes of beings, but our knowledge of that presentation will be limited to the material and conceptual tools at our disposal, including our very concepts of *knowledge* and *writing*.)

Photography as "drawing with light" has been connected intrinsically with inscription from its early days. Yet the actual process of making a mark on the surface was originally seen as a function of a nonhuman agent. In the aptly titled *The Pencil of Nature,* one of the first commercially available books of photography, published between 1844 and 1846, W. H. Fox Talbot writes that "the plates of this work have been obtained by the mere action of Light upon sensitive paper. . . . They are impressed by Nature's hand."[13] The leaving of marks can be cultivated, but, as Talbot himself discovered in the context of his self-avowed inability to draw well, Nature beats the human in precision stakes. As illustrated by the examples of images of landscapes, people, architecture, and still lifes included in *The Pencil of Nature,* the eye of the camera is more accurate and more powerful than the human eye.[14] Talbot was very much aware of the fragility of the resultant photographic recording—indeed, he put much effort into trying to reduce this fragility and fix an image on paper for a prolonged period of time. But the very idea of light acting as "the pencil of Nature," making semipermanent inscriptions to be detected and interpreted by others, positions photography in its nascent state alongside the then newly emergent discipline of geology (with Lyell's *Principles of Geology* having been published between 1830 and 1833), resulting in photographs being seen as thin fossils.

Let us dwell for a moment on this link between photography and geology as different forms of temporal impression by turning to the work of William Jerome Harrison, a nineteenth-century English scientist, teacher, and writer. Harrison authored two seemingly unrelated volumes that reflected his personal and professional interests: *A Sketch of the Geology of Leicestershire and Rutland* (1877) and *History of Photography* (1887).[15] In his intricate reading of Harrison's work, Adam Bobbette shows how, for Harrison, "photography and geology are constituted by similar processes."[16] Full of painstaking detail but rather tame conceptually, Harrison's prose in *History of Photography* occasionally rises to quasi-sublime heights in an attempt to say something bigger about the material at hand. And thus, among the thorough accounts of the different methodologies of the dry and wet photographic processes, Harrison boldly pronounces that "there is nothing new under the sun—especially in photography."[17] This link is not just metaphorical:

> Harrison characterizes the protagonists of the art form as apprentices of impressions. According to his assessment, "impressioning" is a process as ancient as the tanning of human skin under the sun, or the bleaching of wax by the sun. In each case, the sun has created an impression on a body. For Harrison, this was the earliest and most basic form of photography.[18]

Specifically, Harrison looks at the working processes of one of the many simultaneous inventors of photography, Nicéphore Niépce. Niépce's account of photography (called heliography, or "sun-writing")—in which light "acts chemically upon bodies," "solidifies them even; and renders them more or less insoluble"—provides another link between photography and fossilization.[19] This link becomes even more evident once we take into account that "Niépce studied lithographic forms of image reproduction, the geological implications of which are evident: *litho* [*sic*] is Greek for stone, and *lithography* is the process of imprinting an image onto a stone.... Niépce considered, radically, that light could be substituted for human labour as the agent for copying images into stone."[20] We could therefore conclude that, for Harrison, the history of photography is literally a geological history—while Harrison himself becomes the first narrator of the nonhuman history of photography. In the closing words of *History of Photography,* he considers the photographic process to be part and parcel of the geological history of Earth, pointing out "how beautifully it exemplifies the theory of evolution, process rising out of process."[21] Via its link with fossils, photography reveals itself also to be coupled with extinction. In making this link, not only does Harrison pinpoint the nonhuman element of the photographic inscription but he also seems to intimate that photography *has always been there,* in cosmic deep time. It "just" needed to be discovered and then fixed for a little longer—rather than invented. If photography and fossilization are both practices of "the impression of softer organisms onto harder geological forms," photography is not a new process but, instead, a "modern, mediated extension" of the ancient-long "impressioning" activity enabled by light, soil, and various minerals.[22] The human element comes into the picture, literally, as the "apprentice to impressions enabled by the technical-material apparatus of the camera, plate, chemicals and light."[23]

This step into deep time—on Harrison's part, but also on my part here—is an attempt to go beyond the history of photography as part of human history, one that is driven primarily by human motivations and needs. André Bazin's argument, in which photography, together with other plastic arts, is linked with the "practice of embalming the dead" as a way of achieving victory over time, can be seen as the key representative of the humanist approach.[24] Even though something resembling geological vocabulary of "impression" enters Bazin's discourse, with photography being described as "a kind of decal or transfer" resembling "a fingerprint" and contributing "something to the order of natural creation," the process is ultimately linked by Bazin to the human's "deep need" to have the last word in the argument with death.[25] It is precisely this understanding of photography both as an attempt to overcome death and as a constant reminder of it that has set the tone for the discourse on this medium in the twentieth century. No text made this link more explicit and conserved it for future scholars of photography more strongly than the celebrated *Camera Lucida* by Roland Barthes.[26] Barthes's slim volume is a meditation on the death of his mother, prompted by seeing a photograph of her, and, more broadly, on images as affective devices that become placeholders for melancholia and mourning. Yet this narrative, as well as the very choice of images in *Camera Lucida,* ends up confining photography to a permanent struggle against death. The photographic medium thus becomes a memory aid and a mausoleum, with life preserved as a death mask. An attempt to tell a "deep history" of photography as part of the history of Earth the way I have attempted to do it here, via the work of Harrison, a history that transcends human desires and needs, can therefore allow us to outline a different approach to the photographic medium and process. If we recognize that Earth is "a source of invention through the entanglements of form and matter," while the sun is a source of energy and ultimately life on this earth, we can read photography as partaking of their vibrant and life-giving (rather than just life-conserving) properties.[27]

From the perspective of cosmic time, fossilization can therefore be seen not just as the preservation of life but also as the transmission of its evolutionary principle, with all the nonlinear unpredictability and diversity it entails. It is therefore perhaps apposite to try, together with Claire Cole-brook, to "imagine a species, after humans, 'reading' our planet and its archive: if they encounter human texts (ranging from books to machines to fossil records) how might new views or theories open up?"[28] One attempt to envisage such an archive was recently undertaken by photographer Hiroshi Sugimoto in the exhibition *Lost Human Genetic Archive* held at the Palais de Tokyo in Paris in 2014. The concept of fossilization underpins the whole of Sugimoto's oeuvre. "Fossils work almost the same way as photography: as a record of history. . . . To me, photography functions as a fossilization of time," he says.[29] But in Sugimoto's images, fossilization as a way of recording time becomes more than just a figure of speech. As explained in a leaflet accompanying his show called *Still Life* (London, 2014), the fossil is both a historical fact and a photographic conceit for the artist, serving as "a living record and point of departure into history, crystalizing a moment in time into a singular object."[30]

Lost Human Genetic Archive (Figure 3.2) dispels with the minimalism and visual elegance of Sugimoto's photographic projects, such as *Theatres* or *Dioramas,* by presenting the visitor with a rich and diverse collection of objects amassed by the artist over time and arranged into a series of tableaux.[31] Placed somewhere between Dante's Inferno and an enormous toy shop—there are both Barbie dolls and sex dolls on display!—the cavernous basement of Palais de Tokyo presented a *Wunderkammer* for the age of extinction in which new things popped up from around every crevice and corner, on their way to go out with a bang. Each little room in the show staged an alternative "just after the extinction" scenario. Yes, we are going to die, Sugimoto-the-roguish-curator-of-doom seemed to want to remind us, but what a blast we've had—and yet, he frowned, look what a mess we've made. This dual emotion of playfulness and melan-cholia was conveyed by the conflicting visuality of the setup: the visitors wandered around corrugated metal mazelike structures, to be presented constantly with amazing objects: fossils from the Cambrian to the Eocene, one of Sugimoto's own *Seascapes,* astronauts' poo. In what was perhaps a variation on the established trope of the sublime in art, whereby the work

home > *exhibition* > *aujourd'hui, le monde est mort [lost human genetic archive]*

© Vue de l'exposition de Hiroshi Sugimoto « Aujourd'hui le monde est mort » (25.04.14 –
07.09.14), Palais de Tokyo. © Photo : André Morin

Figure 3.2. Website of Palais de Tokyo promoting Hiroshi Sugimoto's exhibition
Lost Human Genetic Archive, 2014. Screenshot by Joanna Zylinska.

evokes pleasure *and* pain, for aesthetic as well as moral effect, Sugimoto
mischievously declared, "Imagining the worst conceivable tomorrows
gives me tremendous pleasure."[32] Yet the exploratory fun offered by the
artist to the visitors to this Lunapark of the Anthropocene carried a serious message: "Where is this human race heading, incapable of preventing
itself from being destroyed in the name of unchecked growth?"[33]

   Minneapolis-based artist Alexa Horochowski's visually restrained
project *Club Disminución* (Club of diminishing returns), instigated during her residency at Casa Poli in Chile, offers an interesting counterpart
to Sugimoto's opulent archive. *Club Disminución* takes to the task the
modernist dream of an ideal society that was to be achieved thanks to
developments in technology and engineering. Casa Poli, a minimalist
cement cube, stands on a jagged cliff overlooking the Pacific, both fore-
grounding and suspending the differentiation between the (hu)man made

and what this human calls "nature." By blurring the boundaries between development and evolution, Horochowski has opened a rift in the modernist narrative of the seamless unification between citizens and their environment, with the promised Corbusierian order resulting in a haunted space, exposed to the elements. Staying in this modernist masterpiece, literally perched at the edge of the world, the artist polluted its visual and conceptual purity with the objects and materials found outdoors: rubbish, fossils, kelp. It is the latter material that provides a conceptual thread for the *Club Disminución* show, first staged in 2014 at the Soap Factory in Minneapolis—one of the many places of traditional industrial production now regenerated into "cultural industries" zones. Kelp, or *cochayuyo,* as it is known in Chile, is a shore seaweed that resembles a thick cable and that arranges itself into unusual quasisculptural tangles. Attracted to its rubbery texture and its strange beauty, Horochowski started collecting the plant in large amounts during the early stages of the project and then hung it in various places in the white cube of Casa Poli. Cochayuyo as a plant that could easily pass for a technological object thus became an inspiration for her to interrogate the intertwining of nature and culture, extinction and obsolescence.

The visitors to Horochowki's exhibition in Minneapolis were greeted by large-screen videos showing this cablelike product, with images cut and mirrored on-screen to form a kaleidoscope of poetic movement. These were accompanied by another playful take on modernist visual art: cubelike structures that may have been made of kelp, cable, or metal wire. Even if we touched them, we were not entirely sure. The displayed objects arranged themselves into what the artist herself has described as "a post-human natural history of the future," whereby "a fossil of a credit card [one of the most intriguing objects in the show!] heralds a post-consumer future, beyond the era of the Anthropocene" (Figure 3.3).[34] This latter artifact raises an intriguing question: what will future generations make of the fossils of those small embellished plastic rectangles that the humans of the late capitalism era have endowed with so much value? Yet Horochowski offers us more than the familiar lamentation over the passing of man and his worldly wealth. As Christina Schmid has put it, the artist's "more-than-vaguely vaginal imagery suggests a gendering of

Figure 3.3. Alexa Horochowski, from *Club Disminución,* 2014. Courtesy of the artist.

the dialogue: it challenges macho modernism's tragic-heroic quest for mastery"—so evident in the jeremiads by the various Anthropocene-era male prophets of doom and gloom who seem to take delight in pronouncing our imminent death.[35]

Club Disminución, in which "the diminishing returns" also refers to seemingly pointless activities, such as straightening kelp, drying it, and fitting it into cuboid shapes, envisages a future beyond the human. It thus becomes a quintessential example of "art after the human," still appreciated "as art" from our human position of here and now, yet appreciated precisely for its placement against the horizon of extinction. In its playful reflection on the passage of time, the work seems to be "encouraging the United States to join the club of formerly great nations and have-beens who lament the waning of past glories." Yet Schmid insists—and I would agree with her—that "*Club Disminución* is not a depressing show," although it may be a melancholic one. "Calmly and not without humor . . . , Horochowski proposes that we dare look into the dark. We can face the sunset, her work argues. *Club Disminución* gathers in the fading light and dwells, affectionately, in the lengthening shadows of the human age."[36]

So, if we humans can never face the sun, what does it mean for us to be able to face the sunset? This question has been addressed, although from a slightly different vantage point, by photo-artist Penelope Umbrico, perhaps best known for her large-scale project *Suns (from Sunsets) from Flickr*. Began in 2006 (when "sunset" was the most tagged word on the image hosting website Flickr), the project explores ideas of originality and replication in the culture of online sharing. The artist zooms in on a snapshot she finds that features a sunset, cuts out the sun from it, resizes it, and adds it to the ever growing grid of burned-out white globes placed against an orange-red background (Figure 3.4). These sun tapestries are then displayed as large printouts on gallery walls but also return to the Internet in different guises—as small grids, a screensaver, a set of virtual postcards.

It is therefore not the banal visuality of the sunset but rather participation in the collective practice of sharing something you cannot claim authorship over that is of principal interest to Umbrico. Yet she also admits to being interested in the sun as the light source, and the transformations of this light source, both on the level of image and matter. Alongside her exploration of the digital environments, Umbrico's concerns are aligned with the traditional perception of photography as a practice of drawing with light, and with the energetic transformations its geological actions undergo on the Internet. In what sounds like a playful rebuttal of the more solemn tenor of certain philosophical propositions about the death of the sun, Umbrico pronounces that "the sun is dead but we make our own light"—and then goes off to rephotograph the suns from Flickr as displayed on the screen with her iPhone and to explore the new light effects produced in the process.[37] The result is a follow-up project, *Sun/Screen* (2014), in which sunset-like hues merge with a moiré pattern caused by the superimposition of the pixel grids, meshes, and dot patterns upon an image. The image then emits an uncannily beautiful light, which does not belong to the sun anymore but which is not entirely *ours* either. Yet our human perception, with its specific visual apparatus and its color recognition capabilities, is required to acknowledge this very denaturalization of the sun into a moiré pattern. In other words, the denaturalized sun needs the human body to experience this "denaturalization."

Figure 3.4. Penelope Umbrico, *Suns (from Sunsets) from Flickr,* 2006-. Screenshot by Joanna Zylinska.

Umbrico's playful projects can be seen as an unwitting response to the philosophical problem posed by Jean-François Lyotard in his essay "Can Thought Go on without a Body?," first published in 1987 and included in *The Inhuman.* Lyotard declares there that the "sole serious question facing humanity today" is the solar explosion that awaits us in 4.5 billion years, as a result of which "everything" will come to an end. The sun's death presents itself to us as the ultimate event of extinction and thus as the ultimate sublime. As Lyotard points out, "after the sun's death there won't be a thought to know that its death took place."[38] Yet the universe, of course, *will* "know" about this death; it will "see," "record," and no doubt "respond" to it, but in a language that far exceeds human communication models and structures. Having outlined this somewhat bleak yet still rather remote prospect of the total annihilation of life, Lyotard then proceeds to mock the efforts undertaken by the cyberneticists of technocapitalism to "make thinking materially possible after the change in the condition of matter" by shifting life to other galaxies to liberate it from the throes of the dying sun. This process seems desperately grotesque as, for Lyotard, embodiment constitutes the very condition of thought: our "software"—mind, philosophy, language—is codependent on, that is, constituted by and constitutive of, our hardware (the body). Even if embodiment is seen as epiphenomenal rather than necessary in the development of the human, Lyotard recognizes that rather than entertain fantasies of extricating human intellect from its material shell, we would be better off getting to the bottom of the desexualized yet so-very-gendered dream of disembodied posthuman thought. From this point of view, the actual disaster that should concern us involves the disappearance not of the solar system as our matrix of reference but rather of the body, that is, the extinction of the human as we know it—*while we are still around.* Accusing philosophers of extricating matter from their writings, Lyotard reminds us that the materiality of the human and of the universe needs to be read alongside its technicity, with matter being taken "as an arrangement of energy created, destroyed and recreated over and over again, endlessly."[39] He pinpoints that

> as anthropologists and biologists admit, even the simplest life forms, infusoria (tiny algae synthesized by light at the edges of tidepools [now termed Protista] a few million years ago are already technical devices. Any material system is technological if it filters information useful to its survival, if it memorizes and processes that information . . . that is, if it intervenes on and impacts its environment, so as to assure its perpetuation at least.[40]

Positioning the emergence of life in early microorganisms as a technical process, Lyotard goes beyond the humanist logic of originary technicity that shaped the work of his contemporaries, such as Bernard Stiegler, whereby it is the human that is seen as constituted by, and emerging with, technicity. For Lyotard, technicity is already the condition and driving force of primordial life. Picking up on this idea, I want to suggest that the process of the emergence of life also reveals itself to be inherently photographic, with light being needed to initiate photosynthesis, that is, to make a lasting change on an organism, and then triggering off further changes. Yet, even if we continue pursuing this expanded understanding of photography as a nonhuman process that exceeds human acts involving cameras and photosensitive material, we are nevertheless returned here, with Lyotard, to the phenomenological experience of light cast upon a human body, located on Earth, which is being lit by its middle-aged sun. Indeed, for Lyotard, corporeality is the condition of knowledge but also of the phenomenological experience that enables openness, generativity, and generosity—and that allows for the transmutation of the technical action of information transfer into an ethical act. This returns us to the issue of the human's inability to face the sun yet still having to take on the task of facing the sunset. The death of the sun, the universe, and ourselves is thus repositioned here from an ontological to an ethicopolitical problem. It is because being able to face the sunset also means coming to terms with the problem of energy—and of the depleting resources not just from the solar domain but also from the terrestrial (or, more specifically, subterranean) realm. Being able to *share* the sunset, in the Umbrico manner, hints at the possibility of thinking—even if not yet actually implementing—a more generous, less exploitative mode of engaging with those resources.

It is precisely the *(mis)management of energy sources as fossil fuels* by the human that is referenced as one of the symptoms of the Anthropocene, a state of events that has resulted in the change of the composition of the atmosphere—and thus in the alteration of the nature of light that reaches us through it. As Kolbert writes, citing the chemist Paul Crutzen (who is credited with popularizing the term *Anthropocene*), "owing to a combination of fossil fuel combustion and deforestation, the concentration of carbon dioxide in the air has risen by forty percent over the last two centuries, while the concentration of methane, an even more potent greenhouse gas, has more than doubled."[41] Facing the sunset is therefore a way of suspending what Finnish philosopher Tere Vadén has called "fossil sense": an assumption that, because things have been a certain way for the last 150 years—with the intertwined logic of economic growth and fossil fuel exploitation shaping our modern ways of life—they will always be this way.[42] Fossil sense is therefore actually *non*-sense: it involves a forgetting of the deep time of history, fueled by myopic self-interest and species narcissism. Vadén claims that our lives are so intertwined with fossil fuels that "our desires are the desires of oil, our dreams the dreams of coal."[43] The modern human is therefore himself fossil fueled, with the very core of not just his physical but also his economic and sociopolitical identity being shaped by hydrocarbons. To extend the concept and metaphor we have been working with here, we can go so far as to suggest that, while our own bodies are made of the (same) star stuff, they now also carry a record of industrially processed hydrocarbons: shards of coal, remnants of oil. Or, indeed, that we ourselves have become a photograph—and a fossil—of our way of life. As Antti Salminen puts it in his poignantly titled essay "Photography in the Age of Fossil Nihilism," "when a life form lives off fossil fuel, it will gradually become fossilized itself."[44]

Living under the cloud of oil fumes and global pollution, we seem to have forgotten about the sun. Grand as this proposition may sound, I want to suggest that photography can be mobilized to address the present fossil crisis in two ways: by expanding the temporal perspective from which this issue is normally seen (or *not seen,* as the case may be) and by helping us outline a different, less deadly solar economy. Some claim that it can most easily undertake this task by serving as a record of the terrible damage done to the environment. We can reference here, for example, the series called *Oil* by Edward Burtynsky, which features large-scale images of oil fields in Azerbaijan, the United States, and Canada; discarded or burning tire piles in California; and oil refineries. These predominantly bird's-eye-view images of what, from above and afar, look like digitally enhanced landscape paintings for the HDR age are intermingled with equally large, yet shot face-on, images of ruined car factories. This kind of representational art is important in being able to visualize the environmental destruction and our damaging relationship to various sources of energy, including the sun. Yet there is also a danger that these images will actually perpetuate the act of *forgetting about the sun,* with aesthetics acting as an anesthetic against the urgency of the environmental situation. As the increasing proliferation of images of disaster and suffering in various media testifies, there is no evidence for perception being a trigger for (moral) action. Indeed, sentimentalism or moral outrage aside, visual oversaturation may actually lead to nonaction. Hanski argues that, owing to the way our sensual and cognitive apparatus has evolved, we humans "are only able to perceive a small region of space and a short length of time."[45] We could therefore conclude that evolution has made it impossible for us to *truly* see evolution—and hence also extinction.

This is why we should not overestimate the role of documentary and representational photographs that take environmental issues as their topic. Yet to state this is not to argue for photography's inherent weakness. Indeed, the argument of my article is that photography is a quintessential practice of life, not just in the sense that, today, it records our lives nonstop but also in the deeper philosophical sense of encompassing life as duration through making incisions in it. In other words, *all* photography, with its capacity to capture light and make it act upon surfaces, acts as a cue for the goings-on of deep time, well beyond human control and human existence. Salminen goes so far as to suggest that "in the fossil nihilist age, photography acts as a reminder of the sun"—and thus of life itself.[46]

It is in this idea that the two temporal lines of my chapter—one oriented toward the past, the other toward the future—come together. Photography as an embalmer and a carrier of imprints testifies to the continued existence of solar energy and to its photosynthesis-enabling capabilities. To say this is not to rewrite the traditional narrative about photography as being about life *rather than* death in any straightforward and naive way. Yet, rhetorically placing photography under the horizon of extinction—a horizon under which it has unfolded on a material level since time immemorial—has allowed us to come out on the side of life and to think fossils beyond the currently dominant fossil nihilism. Fossils and/as photographs can therefore be seen as more than just forms of memento mori: they are also ethical injunctions, pointing and reaching out to life, in both its actual and virtual forms. Citing grief counselor and philosopher Thomas Attig, Thom van Dooren writes in his book *Flight Ways: Life and Loss at the Edge of Extinction* that "in choosing to grieve actively, we choose life."[47] This is precisely where photography as a process of fossilization that keeps a record of time becomes an ethical task, a form of countermourning the passage of time by casting light on solar light. By turning and returning to the sun, we can take first steps toward envisaging a new energetics, one that develops a more ethical relationship to fossils as "layers of ancient, non-human death." Photography as an original practice of light, now often undertaken under the glow of electricity as often as under the glow of the sun, can get us to engage with light anew—even though, in its present digital setup, it is also "contingent upon energy borrowed from oil, a light distilled from death."[48]

NOTES

1. I am borrowing the concept of "affective fact" from Brian Massumi. Extinction yields itself to being interpreted through this concept, *not* because it stands in opposition to some *actual* facts (such as climate change, deforestation, or depletion of energy sources), but rather because it affects us mainly as a threat of things to come. Threat, for Massumi, is "affectively self-causing." Brian Massumi, "The Future Birth of the Affective Fact: The Political Ontology of Threat," in *The Affect Theory Reader,* ed. Melissa Gregg and Gregory J. Seigworth (Durham, N.C.: Duke University Press, 2010), 52–54.

2. Elizabeth Kolbert, *The Sixth Extinction: An Unnatural History,* Kindle ed. (New York: Bloomsbury, 2014), chapter 1.

3. Ilkka Hanski, "The World That Became Ruined," *EMBO Reports* 9, no. 1 (2008): S34.

4. See Donna Haraway, "Situated Knowledges: The Science Question in Feminism and the Privilege of Partial Perspective," *Feminist Studies* 14, no. 3 (1988): 575–99.

5. Call for papers, annual conference of the Center for 21st Century Studies: After Extinction, April 30–May 2, 2015, http://www4.uwm.edu/c21 /conferences/2015afterextinction/afterextinction.html.

6. Jan Zalasiewicz is referenced in Kolbert, *The Sixth Extinction,* chapter 5.

7. Ibid.

8. Jussi Parikka, *The Anthrobscene,* Kindle ed. (Minneapolis: University of Minnesota Press, 2014), chapter 3.

9. Ibid., chapter 2.

10. As Claire Colebrook argues, "the fossil record opens a world *for us,* insofar as it allows us to read back from the brain's present to a time before reading." Colebrook, *Death of the PostHuman: Essays on Extinction, Vol. 1* (Ann Arbor, Mich.: Open Humanities Press, 2014), 23.

11. John Durham Peters, "Space, Time, and Communication Theory," *Canadian Journal of Communication* 28, no. 4 (2003), http://www.cjc-online.ca /index.php/journal/article/view/1389/1467.

12. Ibid.

13. W. H. Fox Talbot, "Introductory Remarks," in *The Pencil of Nature* (London: Longman, Brown, Green, and Longmans, 1844), https://www.gutenberg .org/ebooks/33447.

14. "The operator himself discovers on examination, perhaps long afterwards, that he has depicted many things he had no notion of at the time the camera would see plainly where the human eye would find nothing but darkness." Talbot, *Pencil of Nature,* "Plate XIII. Queen's College."

15. W. Jerome Harrison, *A Sketch of the Geology of Leicestershire and Rutland* (William White: Sheffield, 1877); Harrison, *History of Photography* (New York: Scovill Manufacturing, 1887).

16. Adam Bobbette, "Episodes from a History of Scalelessness: William Jerome Harrison and Geological Photography," in *Architecture and the Anthropocene,* ed. Etienne Turpin (Ann Arbor, Mich.: Open Humanities Press, 2013): 51.

17. Harrison, *History of Photography,* 107.

18. Bobbette, "Episodes from a History of Scalelessness," 52.

19. Harrison, *History of Photography,* 19.

20. Ibid., 52.

21. Ibid., 129. The linearity of Harrison's evolutionary narrative reflects the understanding of evolution as logical progression and betterment across time, rather than, as Stanisław Lem put it nearly eighty years later, "a chaotic and illogical designer [that] does not accumulate its own experiences." Lem, *Summa Technologiae,* trans. Joanna Zylinska (Minneapolis: University of Minnesota Press, 2014), 339–40.

22. Bobbette, "Episodes from a History of Scalelessness," 53.

23. Ibid.

24. André Bazin, "The Ontology of the Photographic Image," *Film Quarterly* 13, no. 4 (1960): 4.

25. Ibid., 8.

26. See Roland Barthes, *Camera Lucida* (New York: Hill and Wang, 1981).

27. Bobbette, "Episodes from a History of Scalelessness," 53.

28. Colebrook, *Death of the PostHuman,* 39.

29. Hiroshi Sugimoto, interview, "Memory," *Art21,* season 3, aired 2005, transcript available at http://www.art21.org/texts/hiroshi-sugimoto/interview-hiroshi-sugimoto-tradition. Transcript also available from Archinect Blogs, http://archinect.com/windscreens/to-me-photography-functions-as-a-fossilization-of-time.

30. Pace Gallery, "Hiroshi Sugimoto: Still Life," press release, Pace Gallery, exhibition running November 21, 2014–January 24, 2015, http://www.pacegallery.com/exhibitions/12709/still-life.

31. *Lost Human Genetic Archive* was not the first show in which Sugimoto explored the history of the human against the horizon of deep time. In 2005–6, he staged an exhibition called *History of History* at Japan Society in New York, in which his own photographs were displayed alongside other artifacts: scrolls, wood sculptures, and, most interestingly for us here, fossils. The appearance of fossilized ammonites, trilobites, and sea lilies in the exhibition was most apposite, according to the artist, given that the fossils are "the oldest form of art" and a kind of "pre-photography," providing a genealogy for his art. From the vantage point of deep time, photography can therefore be seen as "the first art, prehistoric, prehumen." Walter Benn Michaels, "Photographs and Fossils," in *Photography Theory,* ed. James Elkins (New York: Routledge, 2007), 431, 432.

32. Cited in Adrian Searle, "Hiroshi Sugimoto: Art for the End of the World," *The Guardian,* May 16, 2014, http://www.theguardian.com/artanddesign/2014/may/16/hiroshi-sugimoto-aujordhui-palais-de-tokyo-paris-exhibition.

33. Palais de Tokyo, description of *Lost Human Genetic Archive* exhibition, http://www.palaisdetokyo.com/en/event/aujourdhui-le-monde-est-mort-lost -human-genetic-archive.

34. Alexa Horochowski, "Club Disminución," *Photomediations Machine,* January 13, 2015, http://photomediationsmachine.net/2015/01/13/club-disminucion/.

35. Christina Schmid, "A Club at the End of the World," MN Artists, October 16, 2014, http://www.mnartists.org/article/club-end-world. For a critique of the masculinism of the dominant discourses of, and debates on, the Anthropocene, see my book *Minimal Ethics for the Anthropocene* (Ann Arbor, Mich.: Open Humanities Press, 2014).

36. Schmid, "A Club at the End of the World."

37. Penelope Umbrico, talk given at the Photographers' Gallery, London, January 16, 2015.

38. Jean-François Lyotard, *The Inhuman,* trans. Geoffrey Bennington and Rachel Bowlby (Cambridge: Polity Press, 1991), 9.

39. Ibid.

40. Ibid., 12.

41. Kolbert, *The Sixth Extinction,* chapter 5.

42. Tere Vadén, "Fossil Sense," *Mustarinda,* Helsinki Photography Biennial Edition HBP14: 99.

43. Ibid., 101.

44. Antti Salminen, "Photography in the Age of Fossil Nihilism," *Mustarinda,* Helsinki Photography Biennial Edition HBP14: 66.

45. Hanski, "World That Became Ruined," S34.

46. Salminen, "Photography in the Age of Fossil Nihilism," 70.

47. Thom van Dooren, *Flight Ways: Life and Loss at the Edge of Extinction* (New York: Columbia University Press, 2014), 144.

48. Salminen, "Photography in the Age of Fossil Nihilism," 69.

4

The Six Extinctions: Visualizing Planetary Ecological Crisis Today

Joseph Masco

The emerging environmental damage of the industrial age offers up rebounding visions of ecological calamity in the twenty-first century. These dangers are not new but rather have been built slowly over decades of human industry, created in the paradoxical pursuit of security, energy, and profit. We find today that the very tools for building a highly globalized modern society are also cumulative ecological dangers, as the unintended effects of industrial activity (across petrochemical, nuclear, and synthetic chemical regimes) produce hazards that exceed our capacity to control, requiring a new assessment of nature, society, and economy. Collectively, people now face industrial dangers that are planetary in scope and operate on vastly different time scales, challenging perception and action while changing the grounds of the political. What should be our collective orientation toward a future that is shifting radically in both its qualities and risks? How do we recalibrate our senses, actions, and expectations as the long-standing modernist assumption of an ever-increasing security through continual technological revolution is replaced by competing visions of precarity and loss? Familiar dangers (nuclear war) and newer ones (a destabilized climate) challenge global governance while also upending foundational notions about technology, modernity, and progress. The nature of the state as a problem-solving apparatus as well as the state of nature as the dominion of life itself are in play at our historical moment in radical ways, not only as conceptual categories but also as a set of embodied relations.

In sum, the ecological future is not what many residents of the industrialized Global North, who for generations have relied on technological innovation to steadily improve everyday comforts, once assumed it would be. Industrial toxicity is shifting Earth systems—across atmosphere, ocean, ice caps, geology, and biosphere—and is on a trajectory to transform the environmental conditions that promoted such spectacular human expansion over the past ten thousand years (a period that saw the invention of agriculture, the written word, the internal combustion engine, the atomic bomb, and the smartphone). This realization is transforming the focus of security from nation-states and international institutions to the envelope of the atmosphere itself. Atmospheric chemistry is politically emerging as a foundational, but highly changeable, support system for life as we know it. Increasingly, environmental problems operate at the planetary scale, which presents the diversity of global cultures and regional economies with a species-level problem for governance. This shift in perspective requires more than new science, engineering, and statecraft; it also requires new imaginaries, new visions of ecological relationally, and a wide-ranging exploration of the codependence of species and Earth systems. As Dipesh Chakrabarty has argued, climate change is a collective danger that invites us to rethink basic aspects of contemporary life.[1] This chapter is therefore ultimately about conceptualization, about how to think on temporal and spatial scales that exceed human senses. For that, we need both science and art, both technical judgments about material conditions and creative efforts to generate new points of orientation for citizens who are increasingly positioned not only as consumers and members of nation-states but also as vulnerable, if hyperactive, Earth dwellers.[2]

In this regard, the concept of the Anthropocene has been a remarkably powerful intervention in the past few years, moving quickly from a formal proposal within the discipline of geology to a wide-ranging transdisciplinary conversation, generating new research programs, journals, seminars, and workshops across Europe, Asia, and North and South

America.[3] In the past few years, barely a month has gone by without a major conference somewhere on the planet addressing the concept. Here is a partial list from 2013–15:

- January 2013: "The Anthropocene Project," House of World Cultures, Berlin
- March 2013: "Bats in the Anthropocene," Third International Berlin Bat Meeting, Berlin
- April 2013: "Living in the Anthropocene," Evergreen State College, Olympia, Washington
- May 2013: "Society in the Anthropocene," School of Geographical Sciences, University of Bristol
- May 2013: "The History and Politics of the Anthropocene," University of Chicago
- May 2013: "Water in the Anthropocene," Global Water Systems Project, Bonn
- June 2013: "Culture and the Anthropocene," Rachel Carson Center, Munich
- January 2014: "Rivers of the Anthropocene," Indiana University, Bloomington
- February 2014: "Encountering the Anthropocene Conference: Role of the Environmental Humanities and Social Sciences," University of Sydney
- March 2014: "Megafauna and Ecosystem Function: From the Pleistocene to the Anthropocene," University of Oxford
- March 2014: "Science, Politics, and Social Natures in the Anthropocene," Rutgers University, New Brunswick, New Jersey
- April 2014: "Anthropocene Feminism," Center for 21st Century Studies, University of Wisconsin–Milwaukee
- May 2014: "Anthropocene: Arts of Living on a Damaged Planet," University of California, Santa Cruz
- June 2014: "Welcome to the Anthropocene: From Global Challenge to Planetary Stewardship," Association for Environmental Studies and Sciences conference, New York

- July 2014: "Access and Allocation in the Anthropocene," Norwich Conference on Earth Systems Governance, University of East Anglia
- August 2014: "Schelling in the Anthropocene: Thinking beyond the Annihilation of Nature," Bard Graduate Center, New York
- November 2014: "The Anthropocene: Cabinet of Curiosities Slam," Environmental Humanities, University of Wisconsin, Madison
- December 2014: "Im/Mortality and In/Finitude in the Anthropocene," Royal Institute of Technology, Stockholm
- March 2015: "Energy Cultures in the Age of the Anthropocene," Obermann Humanities Symposium, University of Iowa, Iowa City
- April 2015: "The Anthropocene: Confronting Global Environmental Change and Hazardous Worlds," EARThS Conference, Washington State University, Pullman
- June 2015: "The Good Anthropocene," Breakthrough Institute, Sausalito, California
- August 2015: "Geographies of the Anthropocene," Royal Geographical Society–Institute of British Geographers Conference, University of Exeter
- October 2015: "People and the Planet in the Anthropocene," Transformations 2015, Stockholm
- October 2015: "Computational Ecologies; Design in the Anthropocene," University of Cincinnati
- November 2015: "How to Think the Anthropocene," Sciences Po and Centre Alexandre Koyre, Paris
- November 2015: "Social–Ecological Dynamics in the Anthropocene," PECS, South Africa
- December 2015: "Democracy and Resilience in the Anthropocene," Canberra Conference, Australia

As one can see, even this partial view of the phenomenon demonstrates that the Anthropocene is now both an era and a qualifier—linking water, air, land, society, culture, the humanities, Schelling, feminisms, megafauna, and bats as Anthropocenic subjects. The term has positive and negative inflections, involving democracy, resilience, annihilations, immortality, computation, and thought itself. Formally, the Anthropocene was introduced by Paul Crutzen and Eugene Stoermer in 2000 to recognize the industrial-age human as a geological force.[4] The professional geological societies are now debating if there is a stratum in the Earth that is such a clear marker of human activity that it could be the basis for declaring a new geological period.[5] Crutzen initially proposed the start of the industrial age as the "golden spike" of the Anthropocene, suggesting that perhaps the first steam engine or coal plant could mark the start of the new epoch, but more recently he has argued that atmospheric nuclear explosions have left the clearest industrial signature in geology and biosphere. The stratigraphic associations may render a judgment on geological periodization, perhaps elevating the nuclear age to a geological period. What happens to the Cold War when the nuclear age becomes a geological period? Are we prepared to see technological effects so radically decontextualized from their historical and political context, liquidating political epochs in favor of geological time? Or, the geologists might alternatively continue this terrific mischief by postponing their judgment altogether or deciding we have not yet entered a new geological era. No matter how the formal judgments about the Anthropocene are resolved, however, this intervention into contemporary politics has been a bold and brilliant bit of agitprop on behalf of environmental sustainability.

The remarkable discursive success of the Anthropocene in a few short years has produced its share of critics as well. The concept clearly contributes to a scientific discourse within the earth sciences but is somewhat less helpful I think for those disciplines, including the social sciences and humanities, that do not think on geological time scales. The fast adoption of the concept across disciplines raises a series of concerns. Let me briefly consider four hesitations.

First, in the act of recognizing the unintended cumulative consequences of human industrial activity, the term can appear to name people as the core agents on Earth. If we were to limit the Anthropocene to the production of industrial toxicity, I would agree. But there are many kinds of agents on Earth, and it is a mistake to encourage, even as a political exercise or public mobilization strategy, a perception that people are the exclusive or primary actors within ecological systems. In recognizing the cumulative force of human activity on Earth systems, the "Anthropocene" risks creating a metahuman agency, one that fits easily within the very neoliberal worldview that has accelerated extraction and consumption regimes on the planet. The "end of nature" and "age of man" conceptual frames all too easily morph into a nonscientific, popular claim of sovereignty over the Earth rather than underscoring the unintended destructive effects of people upon it, as the authors of the term have hoped it would.

Second, the Anthropocene can easily be constituted as a mirror to the Cold War logics of closed systems, of limited systems interacting in positive and negative feedback loops and thus subject to command-and-control reasoning.[6] This risks reinstalling a kind of anthropocentrism as the Earth's atmosphere has changed dramatically over the eons and has only in the Holocene become conducive for human life and expansion. As shorthand for extremely complex articulations across domains, the Anthropocene installs assumptions of a "normative" planetary state, one that (in its focus on human creaturely comforts) could just as easily in the long history of the planet be considered extraordinary. That is, the Holocene can also be approached as an exceptional era for atmospheric chemistry on Earth, just one particularly useful to the human species.

Third, there are also many societies on Earth that are not particularly Anthropocenic, meaning that the "Anthropocene" is less accurate than talking about specific Anthropocenic societies, economies, and activities. It is vitally important that, in the effort to address planetary-scale ecological change, global inequalities are not subsumed into a species-level critique.[7] The Intergovernmental Panel on Climate Change (IPCC) has worked to address this point directly in its major publications, but the inherent link between planetary-scale activities, geological time scales, and species thinking that is at the center of the concept of the Anthropocene blurs historical distinctions in modes of living and specific concepts of nature.[8] These alternative visions and lifestyles are important not just to recognize but to actively learn from for their collective insights.

And finally, the Anthropocene to date has been mobilized via apocalyptic visions of the future, drawing on tropes developed most directly by nuclear crisis as a tool of political mobilization. Thus we have a language of ultimate crisis designed for one technological problem being used in a context that is not a parallel situation at all. Nuclear war is short and fast and in the hands of a few; climate change is long and slow, a cumulative and accelerating effect of industrial activity. Depictions of collective danger should acknowledge these profound distinctions and the different modes of governance they demand.[9]

Donna Haraway has recently critiqued the Anthropocene, suggesting that it naturalizes a specific historical–political formation—capitalism—as the only human mode.[10] She suggests, along with Jason Moore, that instead of *Anthropocene,* it should be *Capitalocene*—to mark the specifically destructive qualities of a petrochemical-based capitalist system, or perhaps the *Chthulucene,* after the many-tentacled fictional creature created by horror writer H. P. Lovecraft.[11] Bruno Latour has argued that *Gaia,* building on James Lovelock's original proposal, is a more appropriate concept, as it suggests a system with multiple kinds of nonhuman agency, one that is also open ended.[12] That is, the interaction between living beings and Earth systems is not a closed, homeostatic, cybernetic system but one that can take an infinite number of forms at the planetary scale (some suitable for humans, many not). For Latour, what we need now to consciously produce is "geostory"—that is, histories and ethnographies of human activities as geological forces mobilized to create new collectives committed to managing environmental effects. Nigel Clark, following a similar line of thought, has productively proposed a foundational rethinking of the term *geopolitics,* provocatively suggesting that we need now to attend not only to international relations but also to the material conditions of life on planet Earth.[13] I have suggested that we are entering the "Age of Fallout," as environmental crisis is largely the ongoing aftermath of twentieth-century industrialism, raising important questions about temporal lag; environmental perceptions across petrochemical, synthetic chemical, and nuclear regimes; and the cumulative force of technological revolution.[14]

In any case, the necessary core project of reducing toxic emissions requires a public mobilization to deal with a highly complex future danger that engages the total environment. This makes the problem of environmental crisis one not only of science and simulation but also of communication and visualization.[15] How, indeed, can we take infinitely complex

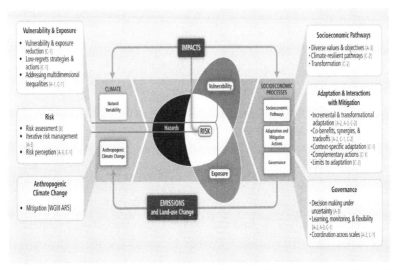

Figure 4.1. The IPCC "Solution Space" schematic illustrates the problem of climate change. Courtesy of the Intergovernmental Panel on Climate Change, *Climate Change 2014, Part A* (2014), 26.

processes involving the interaction of people, industry, atmosphere, water, land, and climate and make them intelligible to non-earth scientists in a way that promotes radical changes in our politics and economy? On this point, it is easy to turn to the IPCC, which, since 1988, has been studying the terms and possible futures of a warming planet. The IPCC is a remarkable scientific and political achievement, as it brings together thousands of scientists from all over the world in an effort to create a consensus view of changes in Earth systems today while also offering scenarios for future climatic changes. Figure 4.1 illustrates how the IPCC visually portrays the problem in its 2014 report.[16]

This is a schematic slide, depicting the complex, interdependent processes discussed in the broader IPCC report. Loaded with technical terms, it is a risk assessment of astonishing complexity combining the total environment with human economic activity and politics. Its central terms—risk, vulnerability, hazards, and exposure—each has substantial literatures across scientific, regulatory, and social theory projects, a complexity only amplified by the use of "socioeconomic processes" as a catchall phrase for human activities (across nations, cultures, and

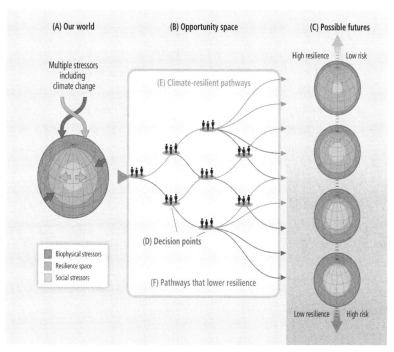

Figure 4.2. The IPCC "Opportunity Space and Climate-Resilient Pathways" illustration showing the human species's relationship to the future. Courtesy of the Intergovernmental Panel on Climate Change, *Climate Change 2014, Part A* (2014), 29.

economies) on planet Earth. The programmatic clarity of this kind of illustration obscures the complexity of its production and cannot include the potential negative outcomes of mitigation schemes or the data friction between component parts.[17] Figure 4.2 presents the IPCC's basic argument about the human species relationship to the future, making the simple, if vitally important, point that what people do today impacts conditions in the coming decades and centuries.[18] It recruits readers to an idea of a resilient global population, one that both mitigates and withstands ecological change. The illustration lends itself to an ideal of global governance without stating who can or will make decisions on behalf of the human species. Thus its species view is at odds with our current political reality, in which the nation-state and corporate forms with their much narrower field of interests—population security and profit—define the contours of the political. And as Brad Evans and Julian Reid have pointed out, the current logics of "resilience" need to be carefully articulated to avoid being a code for social abandonment.[19]

In terms of mobilizing publics to consider existential crisis, the nuclear age offers the prime example. What was called *civil defense* in the United States was a multigenerational effort to teach Americans to fear the bomb in specific ways and therefore mobilize them as nuclear subjects.[20] The mushroom cloud became the emblem of collective disaster after 1945 in the United States, an image that could be confidently evoked to produce a set of cultural associations that were nationally meaningful (Figure 4.3). The mushroom cloud, however, was an explicit branding of the nuclear danger by the U.S. security state, a carefully calibrated image-form that served as an icon of ultimate destruction but that also created a distant viewer, one removed from the event as spectator to it. Nuclear imagery was carefully controlled and circulated during the early Cold War in the United States; it was officially aimed at mobilizing Americans as Cold Warriors. The specific visual tropes of U.S. nuclear culture also created for many an experience of nuclear sublimity—a perverse new kind of attraction to witnessing a destruction that did not need to be felt or shared—because one was visually positioned as external to it. Nuclear imagery thus raises the basic theoretical questions about perception, violence, and collective death that also need to be addressed by contemporary climate politics, which often rely on the tropes of nuclear war—a total destruction—to constitute urgency around carbon emissions and a destabilizing environment. Indeed, a world that is increasingly more hostile to live in (across health, food production, air quality, and weather) is much more difficult to articulate in a single image, as it represents a decreasing quality of life rather than the total destruction of it.

In the immediate aftermath of World War I, Freud explored how death was managed psychically, contemplating how the "most civilized" nations—that is, the most industrialized—proved to be the very ones capable of violence on a new scale. Climate change makes this argument anew, as it recognizes that petrochemical toxicity (in all its forms) is a unique achievement of industrialized society. Freud argued that people maintain two opposed impulses concerning death. The first is a denial

Figure 4.3. Mushroom cloud from U.S. thermonuclear detonation at Bikini Atoll, March 26, 1954. Courtesy of National Nuclear Security Administration/Nevada Site Office.

of one's actual death. Noting the removal of death from public life in Europe at the end of the nineteenth century—the effort to render sickness and death socially invisible—that preceded the invention of world war, Freud wrote,

> We cannot, indeed, imagine our own death; whenever we try to do so we find that we survive ourselves as spectators. The school of psychoanalysis could thus assert that at bottom no one believes in his own death, which amounts to saying: in the unconscious every one of us is convinced of his immortality.[21]

We survive ourselves as spectators. This is a curious phrase, one that underscores the basic tension between an internal expectation of immortality and the conscious knowledge that everyone has a defined life course ending in death. It suggests an internal dissociation toward death, allowing us to be external witnesses to our own injury or end. This contradiction is managed in at least two ways: the first is simply to psychically locate death elsewhere. Freud talks about the power of fiction, of distant wars and disasters, of newspapers, to give a precise location to death, one that allows finitude to be contemplated but also pushed away from the ego and externalized. The other mechanism is what Freud would eventually call the *death drive,* a basic orientation of the human organism toward an ultimate release from physical struggle and a return to a state of inertness.[22] The pursuit of pleasure and gratification may motivate life, but a drive toward a feeling of nothingness is coterminous with it. Thus, for Freud, we all have a complex orientation toward our own death, one that is managed via displacements and overdeterminations but that is also linked to an internal desire for an ultimate release, an embrace of a pain-free oblivion.

This raises an interesting question concerning planetary-scale existential danger: when we contemplate images of the end, of the apocalypse, of extinction, what kind of work are we doing? Is the pleasure in that mode of reflection an orientation toward the last and final release, or is it alternatively the deflection that allows the ego to feel removed from death and, if not immortal, then at least served and entertained by locating such finality elsewhere? Think about this next time you go to the movies and are confronted by a story of nonstop mayhem or read of a cataclysm on the other side of the world with interested detachment. The aesthetics, even erotics, of death is a long-running cultural concern in Western theory but one that today intersects with an emerging planetary consciousness, one that demands scaling local dangers up to the earthly sphere and back again. Dangers are revealed today as simultaneously hyperlocal formations and as planetary concerns, a fact that requires an evolutionary collective consciousness, one merging a new orientation toward consumer desire with a new kind of statecraft.

For the rest this chapter, I'd like to explore some dimensions of contemporary ecological crisis, and the broader issue of visualizing collective danger, by engaging a remarkable 2013 exhibition at the Renaissance Society Art Museum in Chicago, curated by Hamza Walker. An explicit engagement with the aesthetic pull of extinction, Walker's *Suicide Narcissus* exhibition helps us think about the limits of human perception as well as the psychosocial effects of radical collective endangerment.[23] Walker presents six artworks, each in a different medium, inviting us to think carefully through questions of collective loss and extinction.

In ancient Greek mythology, as you will remember, Narcissus was a boy of unusual beauty who, failing to return the love of wood and mountain nymphs, was cursed by the god of retribution, Nemesis. The curse, of course, was exquisite. Having never seen his own image, Nemesis leads Narcissus to encounter it in a pool of water. Not recognizing himself, Narcissus falls so powerfully in love with what he sees that he cannot avert his gaze, eventually wasting away by the side of the pool until he dies of starvation. This depiction of a misrecognition with total identification has been a powerful concept for psychoanalysis, informing the logics of Lacan's mirror stage as well as Freud's earlier notion of the ego-ideal, an internalized image of a perfected self that is unattainable as lived experience. The story of Narcissus has come to inform how we think today about self-absorption and ego formation as well as both love and death. But we could also underscore other aspects of the myth less remembered today. It is also fundamentally a story about ecological retribution, as the Earth spirits offended by Narcissus call down a divine and absolute retribution against him, ending in a death that Narcissus could avoid simply by changing his field of vision, simply by looking away. Nemesis's curse offers a terrible justice for creatures of nature injured by the self-absorption and vanity of human beings, a retribution that is played out to the point of an extinction.

Walker's *Suicide Narcissus* offers a rich and varied set of interventions on planetary ecological crisis, the most important but also most conceptually challenging issue of our time. Each artist in the exhibition offers a specific point of view on monumental collective loss, inviting viewers to consider not just the aesthetic forms but also the current conceptual frames available for thinking past one's own existence. Each piece in the exhibition is meticulously crafted and quite clever, working together to create a rebounding provocation about the ability to perceive immanent loss. Contemplating the end, in this case, is not a vehicle for distraction. Precisely because each piece is such a disciplined statement, the exhibition establishes the grounds for cultural critique, a mode of address that can generate a productive shock in the viewing subject.

Walker's intent is to interrogate the conceptual moves that allow one to stand outside an ongoing collective disaster and merely observe it as spectacle. Commenting on the American love of the special effects–driven disaster movie, he writes,

> Global warming and summer blockbusters have been in lockstep, record-breaking temperatures corresponding to record-breaking box office earnings. Draped over summer's Hollywood tent-poles, as these big budget films are called, are plots sagging under the weight of humanity's impending demise, whether it is in the hands of the rabid zombies as in *World War Z,* or whether you happened to be holed-up in James Franco's pad during the rapture as in *This is the End.* The threat of our end is a story as recyclable as cardboard. While ours is certainly not the only story to tell, we are for better or worse the narrator, one whose sense of standing outside the story as it involves our death is a form of denial. The trilobites tell us what we already know, that "happily ever after" is a chapter belonging to another species.[24]

The fatalism in Walker's statement here is, of course, undercut by the sophistication of his curatorial work, which invites us to consider how death as spectacle functions today. *Suicide Narcissus* also offers a variety of vantage points from which to constitute a different collective politics.

In what follows, I'd like to reflect on each of the six artworks in the exhibition, exploring the temporalities, ecologies, and visual logics of total endings under the rubric of the six extinctions.

EXTINCTION 1: LUCY SKAER, *LEVIATHAN'S EDGE*

Lucy Skaer's monumental installation *Leviathan's Edge* (2009, whale skeleton and drywall) offers viewers a compelling figure-depth problem, as a giant white skeleton appears, just on the edge of intelligibility, through the cutaways of a white-walled enclosure (Figures 4.4 and 4.5). The white-on-white context of the installation shifts our perspective as we move closer to an obviously large—too large, in fact; what could it be?—animal that exceeds our field of vision. The partial, carefully framed points of view we are allowed of the skeletal remains draw attention to the partial vision we always have on fossilized life. Embedded in earth or stone, ancient remains are always fragile and partial, requiring some degree of reconstruction and imagination.

Figure 4.4. Lucy Skaer, *Leviathan's Edge,* 2009 (installation view, *Suicide Narcissus,* 2014, Renaissance Society at the University of Chicago). Courtesy of the Renaissance Society.

Figure 4.5. Lucy Skaer, *Leviathan's Edge,* 2009 (sectional view, *Suicide Narcis-sus,* 2014, Renaissance Society at the University of Chicago). Courtesy of the Renaissance Society.

Leviathan's Edge invites us to consider the creatures now long gone as well as the contemporary beings on the brink—the ongoing extinctions—and the value systems people attribute to endangerment.[25] Its partial field of vision requires us to fill in and guess at the creature itself, which also posits the remains as timeless, perhaps ancient, perhaps contemporary, perhaps from the future. It invites us to think of how natural history is enmeshed in human history, while underscoring the fragmented and partial vision any standpoint allows on death itself. The whale skeleton—that might be a dinosaur, or some other fantastic being we do not have a name for—is beautiful and beyond full comprehension here because it remains both mediated and fragmented. It also reminds us that there is not enough space in all our museums to account for all the long-gone life-forms. The small slivers of visual access we have to the remains, the cutouts in the installation or the fossil traces of long-gone life on planet Earth, also underscore our limited ability to apprehend, let alone comprehend, the edges of an extinction, its almost-here-ness or, as Thom Van Dooren might put it, the "flight ways" of ongoing species loss.[26]

Biologists will tell you that over 99 percent of the life-forms that have ever lived on Earth have gone extinct.[27] Extinction is not the exception but rather the rule within the deep history of life on this planet. The best estimates today are that some 4 billion species have evolved over the past 3.5 billion years. In addition to the process of natural selection in eliminating particular species, there have been five mass extinction events, periods when, owing to planetary-scale climatic changes, two-thirds or more of all the organisms on Earth have disappeared. Thus not just species but entire ecosystems die with some regularity: this makes every mode of living both an evolutionary accomplishment and a fragile historical achievement of the first order.[28]

Today, there is much discussion of a sixth mass extinction event, an ongoing shift in the terms of living on Earth drawn from the combined impacts of habitat destruction, pollution, overharvesting, invasive species, and human population growth.[29] This sixth mass extinction will be unique in this planet's history, as it does not arrive in the form of an asteroid collision or volcanic eruption but rather through the hyperactive work of one indigenous species: people. The industrial-age human has become an ecological, even a geological, force, constituting a future of fewer species, reduced biodiversity, and potential disruptions in the food chain.[30] Skaer's *Leviathan's Edge* invites us to consider the once and future remains of monumental life on planet Earth and to consider the "edge," that is, to locate the precise threshold of such a cataclysm, the tipping point between life and nonlife.

EXTINCTION 2: KATIE PATERSON, *ALL THE DEAD STARS*

Katie Paterson's laser-etched *All the Dead Stars* (2009, laser-etched, anodized aluminum) poses the problem of extinction directly, while also raising subtle questions about temporal lag and misperception (Figures 4.6 and 4.7). Documenting the twenty-seven thousand known dead stars in the universe, the installation requires us to think not about galactic space and infinity but about the quality of light from dead stars and the temporality of seeing.

Figure 4.6. Katie Paterson, *All the Dead Stars,* 2009 (installation view, *Suicide Narcissus,* 2014, Renaissance Society at the University of Chicago). Courtesy of the Renaissance Society.

Figure 4.7. Katie Paterson, *All the Dead Stars,* 2009 (installation view, close-up, *Suicide Narcissus,* 2014, Renaissance Society at the University of Chicago). Courtesy of the Renaissance Society.

The speed of light travels at just under three hundred million meters per second. Sunlight takes just over eight minutes to travel from the surface of the sun to the surface of Earth. Thus, when looking at the night sky, there is a lag between the light one sees and its point of origin, meaning that some fraction of the light is from stars that are actually dead, with the last flicker of light energy just now reaching our planet. How, then, with the limited human senses that we have, can we actually see extinction? Under what terms and temporalities does a total loss become visible? Presenting a universe of potential answers to this question in a spherical form evoking Earth, Paterson draws attention to the kind of light our sun gives out, the time that light takes to travel across the solar system, and what it might look like in some five to ten billion years, after it finally goes supernova and burns out.

Paterson also alerts viewers to the fact that extinction is all around us, a constant presence. Thus the challenge is to account for how human senses operate within the ongoing temporal lag between living and dying.[31] *All the Dead Stars* provokes questions about how many species, processes, and ideas are merely the afterimage of themselves—a loss that has already occurred but is just not yet visible as such.[32] Today, pollinators—the bees, moths, and butterflies that enable plants of all kinds to reproduce—are in crisis. The colony collapse disorder among honeybee populations is part of a larger shift in how these insects live and die, how their bodies have been manipulated by people and put to work for agricultural industry.[33] Theories abound, but the likely cause is the combined effects of chemical fertilizers, climate change, and pollution. Put differently, we see in the vast die-offs of pollinators today a variety of species under environmental stress to a kind of maximal degree. Are these a terrestrial version of Paterson's dead stars—a question simply of the temporal lag, not the end result?

One ambitious proposal for climate governance involves determining the operating parameters for Earth systems and mobilizing human society to keep specific domains within the peak thresholds for current life. Known as the planetary boundaries proposal, this entirely reasonable act of environmental governance seeks to transform the climate crisis into "a safe operating space for humanity," to constitute a new form of "planetary stewardship" (Figure 4.8).[34] The idea is breathtaking in its vision, detail-

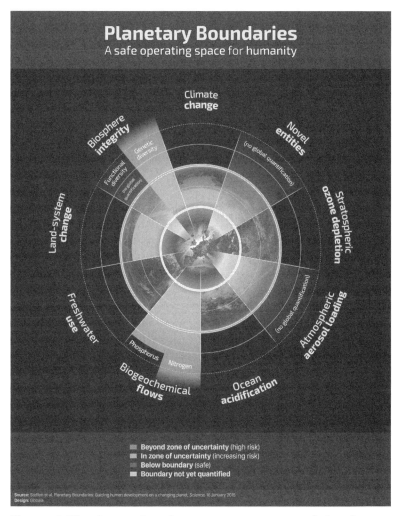

Figure 4.8. The planetary boundary concept presents a set of nine planetary boundaries within which humanity can continue to develop and thrive for generations to come. Stockholm Resilience Centre, *The Planetary Boundaries*. Courtesy of F. Pharand-Deschenes/Globaïa.

ing nine core areas for planetary management—climate change, ocean acidification, ozone depletion, the nitrogen and phosphorus cycles, global freshwater use, land use, biodiversity loss, atmosphere aerosol loading, and chemical pollution. As an aspiration politics, this is a significant intellectual contribution, as it links individuals, organizations, and states to specific goals that can be measured and encourages all parties to engage in planetary thinking. But it is also an engineering approach to planetary process, striving for sublime levels of control of complex human, technological, and ecological interactions. Earth processes are also not closed systems and can be subject to both abrupt and long-term changes.[35] Thus the promise of the planetary boundaries project is to intervene across ecological domains and to calibrate complex planetary systems with multimodal interactions and feedbacks to support specifically human comfort levels. A highly nuanced proposal for the geoengineering of Earth systems, the planetary boundaries process embraces a macroview of environmental risk. Here the call for subtle control of planetary processes overwhelms more simple and direct human-centric responses to climate change—for example, reducing meat consumption, leaving the remaining oil in the ground, reforesting, and committing on a collective scale to renewable energy. The danger here is that the sublime intricacies of Earth systems can become an invitation to a new kind of suicide narcissus—promoting an aestheticized love of complexity and command-and-control reasoning, one that prevents people from simply looking away, seeing a simpler and more direct solution, embracing a different way of living.

EXTINCTION 3: THOMAS BAUMAN, *TAU SLING*

Thomas Bauman's *Tau Sling* (2008, wood rope, motor, mirror) is a mechanized installation in which a thick rope is continuously twisted and reflected in a mirror, constituting an ever-changing, ever-tightening noose (Figure 4.9). It is mesmerizing precisely because of its slow-moving tangle of elements, a machinery that fascinates as it constricts and knots endlessly.

Tau Sling offers a beautiful metaphor for many of today's industrial legacies while also referencing forms of direct (particularly racialized) violence in the Americas. In its ever-twisting form, we can consider how the careful work of building a machine, a security system, an energy infrastructure, a global economy, creates looping and treacherous side effects that can undermine basic logics of safety and sustainability. Bauman's *Tau Sling* is a machinery of insecurity, one constantly in motion,

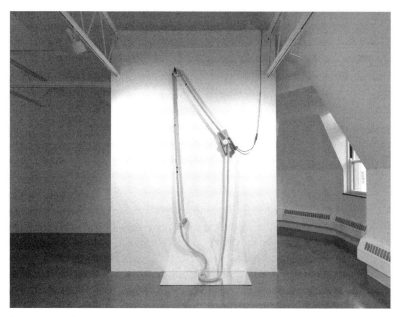

Figure 4.9. Thomas Bauman, *Tau Sling*, 2008 (installation view, *Suicide Narcissus*, 2014, Renaissance Society at the University of Chicago). Courtesy of the Renaissance Society.

offering an enticing, even hypnotic vantage point on constriction and entanglement.

In 1955, mathematician John von Neumann addressed this point directly in a remarkable essay titled "Can We Survive Technology?" in *Fortune* magazine. He writes,

> "The great globe itself" is in a rapidly maturing crisis—a crisis attributable to the fact that the environment in which technological progress must occur has become both undersized and underorganized.... Literally and figuratively, we are running out of room. At long last, we begin to feel the effects of the finite, actual size of the earth in a critical way.[36]

Von Neumann contemplates how technological revolution has expanded to an unprecedented scale, leaving no room for experimentation without collective danger. He focuses on two technologies of immediate concern: nuclear weapons and weather control. The nuclear revolution, he states, will transform "everything it touches," potentially providing new sources of energy, but only if nuclear war can be avoided. Thus surviving the nuclear revolution is the necessary first step before a better world can be built through its peaceful application. Similarly, he considers weather modification as an emerging technology and worries about how carbon emissions from industrial production could produce a substantial warming of the planet, leading to melting ice caps and a changing climate.[37] Industrial emissions, he suggests, could be a planetary problem much like the atomic bomb, constituting a new kind of threat, one exacerbated by the implementation of weather control as a new tool of war. Thus his positive utopia of weather modification is undermined by an imminent new world of "climatic warfare." For von Neumann, technological revolution now affects the entirety of the planet, leaving no future room for expansion and collective costs for failure or error.

One might consider von Neumann's essay a foundational statement in the literature of the Anthropocene, as it suggests that technological growth scaled to the planetary dimension exceeds human understanding and control to become a force in its own right. Peter Haff, following this line of thinking, has proposed that we now live within a "technosphere," a planetary-scale imbrication of technological systems that interact with all Earth systems.[38] Indeed, we have achieved a world of climate alteration, not through conscious decision making or warfare, as von Neumann predicted, but through an unrestrained petrochemical-based capitalism. Von Neumann's fascination with technological revolution and industrial scaling thus works here very much like Bauman's installation—a mesmerizing apparatus of perpetual motion involving an ever-tightening noose, or, as the great mathematician put it in 1955, "for progress there is no cure."[39]

Consider a remarkable image from *The Lancet* (Figure 4.10), which makes a vitally important point about climate change and does so in an entirely new way: it illustrates the proportional local production of carbon emissions in relation to their proportional regional health effects.[40] The planetary politics of carbon here are an emerging violence connecting the Global North to the Global South, producing what Rob Nixon would call the slow violence of anthropogenic illness.[41] But of course, thinking with

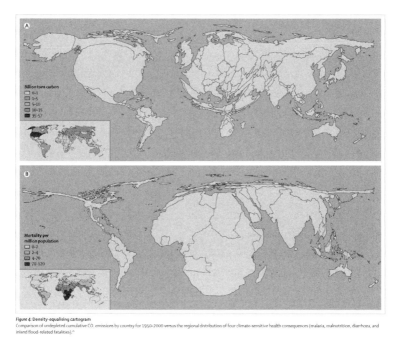

Figure 4: Density-equalising cartogram
Comparison of undepleted cumulative CO. emissions by country for 1950-2000 versus the regional distribution of four climate-sensitive health consequences (malaria, malnutrition, diarrhoea, and inland flood-related fatalities).^

Figure 4.10. Source of carbon emissions to location of increased illnesses from carbon emissions. Reproduced with permission from Anthony Costello et al. and *The Lancet*.

Bauman's installation, carbon is a noose that tightens in all directions, for as the Global South achieves middle-class consumption standards over the coming century, the health effects of an expanding southern petrochemical economy will be directed northward as well—creating an ongoing spiral of negative planetary health effects. Consumption in the Global North entangles health in the Global South, and vice versa, creating an ever more violent circuit that only temporarily offers some the possibility of a detached view on human and nonhuman suffering.

EXTINCTION 4: DANIEL STEEGMANN MANGRANÉ, *16MM*

Daniel Steegmann Mangrané's *16mm* (2008–11, 16mm film shot with a specially modified camera, color, synchronized four-channel digital sound) offers a slow, five-minute tracking shot directly into a rain forest (Figure 4.11). Although the rain forest is seemingly unpopulated, the

Figure 4.11. Still from Daniel Steegmann Mangrané, *16mm*, 2008–11. Courtesy of the artist.

sound track tells us that the space is filled with wildlife unseen. In the *Suicide Narcissus* exhibition, *16mm* was projected in a darkened room against a white screen.

The beautiful canopy of plants and trees revealed here is dense, and there is no trail to follow or trace of a human footstep. Instead, the 16mm camera glides and hovers through the biosphere, moving seamlessly through a jungle space that would be difficult for a person to traverse. What is this point of view the camera offers us? Who are we that we can see in this way? The film documents the archetypal endangered space of our time—rain forest—valued for its intense biodiversity as well as its ability to manage carbon in the atmosphere. But there is also something uncanny about the camera movement (mechanically synced to the rotation of the film spool) that suggests a kind of archival project, one in which the end of the tracking shot could also be the end of the garden itself. The lack of a human trace also allows this image of the biosphere to be both perfect and timeless, a museum of nature for a world that has eliminated such spaces. It is a vision of the forest without us but one carefully crafted to give viewers a privileged, if decidedly nonhuman, point of view.

In Hamza Walker's installation, a cutaway in the projection booth reveals the 16mm projector itself, showing the looping analog film stock as it rattles through the machine. This is an analog photographic technology that is near extinction in the digital age as well as a film about extinction. Here we might consider the multiple ways in which a petrochemical economy threatens the rain forests of our time as well as the medium of film itself as a petrochemical emulsion that structured the social consciousness of the twentieth century. The fossil fuels that we use to run our economy are derived out of the decomposition of plant and animal organisms going back hundreds of millions of years.[42] Fossil fuels are thus, in a very literal sense, congealed time.[43] Film is a petrochemical medium that measures time. Thus Mangrané's *16mm* project can be read as a complex statement on how we perceive and instrumentalize time itself. It also underscores how the emulsions that enable filmic vision participate in the larger petrochemical extraction regime that has wreaked havoc from the polar ice caps to the rain forests to the deserts of the Middle East. These are "chemical regimes of living," to quote Michelle Murphy, that remake society, ecology, and biology in complex and novel formations.[44] That so many of the iconic technologies of modernity from oil to fertilizer to film are also highly destructive petrochemical forms, flammable and unstable over time, reveals at another level the unending challenge of truly valuing, of really seeing, a nonrenewable resource as it is used up.

EXTINCTION 5: NICOLE SIX AND PAUL PETRITSCH, *SPATIAL INTERVENTION I*

Nicole Six and Paul Petritsch's *Spatial Intervention I* (2002, video) is a twenty-eight-minute video, exquisitely photographed, with a simple premise: it presents a man on a frozen lake hacking away at the ice he is standing on until it collapses (Figures 4.12 and 4.13).

The video demonstrates the dangers of undermining the natural systems we depend on for support: ice, atmosphere, and land. It also is a marvelous study of labor in the era of climate change, as the lake here is not so easily broken. It takes twenty-eight minutes of hard, sweaty work to get to the ultimate result. The end, of course, does not happen on screen but with a cut to black and a scream, leaving its final form in our imagination. Extinction here resists representation, becoming something that can

Figure 4.12. Nicole Six and Paul Petritsch, *Spatial Intervention I,* 2002. Courtesy of Bildrecht, Vienna 2015.

only be staged suggestively. In this work, Narcissus sees himself reflected in the frozen lake and just keeps hammering away, transfixed on breaking through to the other side, even if it means his immediate doom. The film, which forces the viewer to attend to each stroke of the ax, creates moments of boredom, suggesting the everyday activities—the unnecessary trip in the car or plane, the purchase of the plastic-wrapped produce shipped from the other side of the world, the eating of the hamburger—that collectively move climate.

Spatial Intervention I also asks us to consider which strokes of the ax matter in the end, which ones do permanent damage.[45] In Walker's curation, the sound of the ax hitting the frozen lake echoes through the

Figure 4.13. Nicole Six and Paul Petritsch, *Spatial Intervention I*, 2002 (installation view, *Suicide Narcissus*, 2014, Renaissance Society at the University of Chicago). Courtesy of the Renaissance Society.

gallery long before one sees Nicole Six and Paul Petritsch's work, setting up a peculiar experience of viewing and hearing alternative takes on extinction. Thus, like many of the pieces in the overall exhibition, *Spatial Intervention I* is ultimately about perception and temporality, and about time running out.

In their professional assessments of the Anthropocene, earth scientists mobilize many forms of data to depict the radical changes that begin around 1950. Population, GDP, water use, paper consumption, telephones, tourism, all take off dramatically after World War II. These indexes are tied directly to parallel appraisals of carbon emissions, ozone depletion, floods, biodiversity loss, and so on.[46] McDonald's restaurants are a key indicator of climate change in these accounts, as cows are a huge contributor to greenhouse gases at every stage in the production of the hamburger. These metrics are crucial, as they recognize how everyday, routine consumer "desire" now constitutes a planetary force. Not coincidentally, the "great acceleration," as many call this shift in global consumption patterns at mid-twentieth century, is also coterminous with the nuclear age—making 1950 the inflection point for both climate crisis and nuclear crisis.[47] How to see everyday activities—eating, transportation, water use, and energy use, as well as national security—as a planetary force is an immediate challenge today. After all, industrial civilization has been largely devoted to creating and expanding creature comforts as the very index of progress since the European Enlightenment. Consumer pleasure is, therefore, at the heart of climate change, requiring not only the complex analytics of the IPCC but also understandings of the historical production of psyches, cultures, desires, and even nervous systems.

EXTINCTION 6: HARIS EPAMINONDA AND DANIEL GUSTAV CRAMER, *THE INFINITE LIBRARY*

The selections Hamza Walker presents from Haris Epaminonda and Daniel Gustav Cramer's *The Infinite Library* (2007 to present, twelve of sixty artist's books) posit the question of knowledge directly (Figure 4.14). Here books, mostly drawn from the mid-twentieth century, have been broken apart and the pages reassembled in radical new combinations. This project of deconstruction and assembly shifts the nature of the human archive, exploding genre and language, to produce new texts, a new collection of works for a post-Enlightenment library. Encased in glass as paired artifacts, *The Infinite Library* attempts a kind of species thinking of the literary, merging fiction, science, self-help, and the arts into a new uber-category. The books explode the evolution of the modern sciences and humanities, a core product of modernity, to offer stories of a new kind, but made from preexisting intellectual and artistic materials.

The question this installation raises for me is, Who is the reader of these texts? Are they imagined as artifacts collected from the ruins by an extraterrestrial archeologist? This postgenre library is one that needs a future reader, perhaps one who does not yet exist. It is an infinite library both because the fragments of existing texts can be endlessly reorganized and also, perhaps, because the very modernist forms of reasoning that have created our current notions of genre are complicit with the

Figure 4.14. Haris Epaminonda and Daniel Gustav Cramer, *The Infinite Library,* 2007 to present (installation view, *Suicide Narcissus,* 2014, Renaissance Society at the University of Chicago). Courtesy of the Renaissance Society.

nation-states and industrial logics that have produced the linked nuclear and climate crises. It took experts of all kinds to build our current archive, just as it took the combined work of physicists, engineers, chemists, and mathematicians to create nuclear weapons and a petrochemical economy. These cumulative knowledge projects implicate the archive itself in a kind of autodestruction on a species scale. Epaminonda and Cramer, however, offer a vision of an alternative archive, one in which the disciplinary lines developed from the French philosophies of the first encyclopedia project (an inaugural event of the Enlightenment) to the drone killing machines circling parts of Earth today are no longer coherent. The accumulated knowledge of humanity is present, but radically reorganized, montaged to a new aesthetics and potentially new outcomes. *The Infinite Library* might well ask, How do we keep the knowledge base of modernity while constituting a different science, technology, and art to enable a different collective future?

ON ENDINGS

In Hamza Walker's curatorial vision, the first step to dealing with collective ecological danger is to become attuned to it at the levels of consumption, image making, and knowledge economies. It is to resist the spectacle of mass death to contemplate alternative futures and become invested in both collective and individual futures. *Suicide Narcissus* attempts to shock the viewer out of a normalized consumer economy, one in which disaster is not a call to collective consciousness but rather a spectacle to be enjoyed via psychic distancing, consumer satisfaction, and depoliticization. Walker asks us to reconsider, and to imagine, a future that operates on vastly different terms, precisely by inviting viewers with such clarity and precision to consider the total cost of not doing so. Thus the immediate answer to the problem of visualizing planetary ecological crisis today is not to consolidate climate change into a single image, offering a mushroom cloud for a new emergency, but rather to proliferate modes of conceptualization and visualization of ecological conditions that can allow wide contemplation of the complexity of human interventions into natural processes and, most importantly, evolve radically with those understandings.

NOTES

My thanks to Michelle Bastian and Thom van Dooren for their engagement and collegiality. Special thanks to Richard Grusin and the Center for 21st Century Studies for their support and editorial care with this chapter.

1. Dipesh Chakrabarty, "The Climate of History: Four Theses," *Critical Inquiry* 35 (2009): 197–222.

2. See Heather Davis and Etienne Turpin, *Art in the Anthropocene* (London: Open Humanities Press, 2015).

3. See Clive Hamilton, Chrisophe Bonneuil, and Francois Gemenne, eds., *The Anthropocene and the Global Environmental Crisis* (New York: Routledge, 2015).

4. Paul J. Crutzen, "The 'Anthropocene,'" *Journal de Physique IV* 12, no. 10 (2002): 1–6.

5. Jan Zalasiewicz, Colin N. Waters, Mark Williams, Anthony D. Barnosky, Alejandro Cearreta, Paul Crutzen, Erle Ellis et al., "When Did the Anthropocene

Begin? A Mid-Twentieth Century Boundary Level Is Stratigraphically Optimal," *Quaternary International* 383 (2014): 196–203.

6. See Paul Edwards, *The Closed World: Computers and the Politics of Discourse in Cold War America* (Cambridge, Mass.: MIT Press, 1996).

7. Christian Parenti, *Tropics of Chaos: Climate Change and the New Geography of Violence* (New York: Nation Books, 2011); Rob Nixon, *Slow Violence and the Environmentalism of the Poor* (Cambridge, Mass.: Harvard University Press, 2011).

8. Intergovernmental Panel on Climate Change, *Climate Change 2014: Impacts, Adaptation, and Vulnerability. Part A: Global and Sectoral Aspects. Contribution of the Working Group II to the Fifth Assessment Report of the Intergovernmental Panel on Climate Change* (Cambridge: Cambridge University Press, 2014).

9. Joseph Masco, "Catastrophe's Apocalypse," in *The Time of Catastrophe: Multidisciplinary Approaches to the Age of Catastrophe,* ed. Christopher Dole, Robert Hayashi, Andrew Poe, Austin Sarat, and Boris Wolfson, 19–46 (Burlington, Vt.: Ashgate, 2015).

10. Donna Haraway, "Anthropocene, Capitalocene, Pantationocene, Chthulucene: Making Kin," *Environmental Humanities* 6, no. 1 (2015): 159–65.

11. Jason W. Moore, *Capitalism in the Web of Life: Ecology and the Accumulation of Capital* (New York: Verso, 2015).

12. Bruno Latour, *Facing Gaia: A New Inquiry into Natural Religion,* Gifford Lectures (2013), http://www.ed.ac.uk/arts-humanities-soc-sci/news-events/lectures/gifford-lectures/archive/series-2012-2013/bruno-latour.

13. Nigel Clark, "Geo-politics and the Disaster of the Anthropocene," *The Sociological Review* 62, no. S1 (2014): 19–37.

14. Joseph Masco, "The Age of Fallout," *History of the Present* 5, no. 2 (2015): 137–68.

15. Eva Lövbrand, Johannes Stripple, and Bo Wiman, "Earth System Governmentality: Reflections on Science in the Anthropocene," *Global Environmental Change* 19 (2009): 7–13.

16. IPCC, *Climate Change 2014,* 26.

17. See Birgit Schneider, "Climate Model Simulation Visualization from a Visual Studies Perspective," *WIREs Climate Change* 3 (2012): 185–93, doi:10.1002/wcc.162; Paul Edwards, *A Vast Machine: Computer Models, Climate Data, and the Politics of Global Warming* (Cambridge, Mass.: MIT Press, 2013).

18. IPCC, *Climate Change 2014,* 29.

19. Brad Evans and Julian Reid, *Resilient Life: The Art of Living Dangerously* (Cambridge: Polity Press, 2014).

20. Joseph Masco, *The Theater of Operations: National Security Affect from the Cold War to the War on Terror* (Durham, N.C.: Duke University Press, 2014).

21. Sigmund Freud, *Reflections on War and Death,* trans. A. A. Brill and Alfred Kuttner (New York: Moffat Yard, 1918), 41.

22. Sigmund Freud, "Beyond the Pleasure Principle," *Standard Edition* 18 (1920): 39.

23. Hamza Walker, *Suicide Narcissus* exhibition essay (Chicago: Renaissance Society, 2013), http://www.renaissancesociety.org/publishing/11/suicidenar cissus/.

24. Ibid.

25. Tim Choy, *Ecologies of Comparison: An Ethnography of Endangerment in Hong Kong* (Durham, N.C.: Duke University Press, 2011).

26. Thom Van Dooren, *Flight Ways: Life and Loss at the Edge of Extinction* (New York: Columbia University Press, 2014).

27. Michael Benton and Richard J. Twitchett, "How to Kill (Almost) All Life: The End-Permian Extinction Event," *Trends in Ecology and Evolution* 18, no. 7 (2003): 358–65.

28. David B. Wake and Vance T. Vredenburg, "Are We in the Midst of the Sixth Mass Extinction? A View from the World of Amphibians," *Proceedings of the National Academy of Sciences of the United States of America* 105 (2008): 11466–73.

29. Elizabeth Kolbert, *The Sixth Extinction: An Unnatural History* (New York: Henry Holt, 2014).

30. Rodolfo Dirzo, Hillary S. Young, Mauro Galetti, Gerardo Ceballos, Nick J. B. Isaac, and Ben Collen, "Defaunation in the Anthropocene," *Science* 345, no. 6195 (2014): 401–6.

31. Douglas J. McCauley, Malin L. Pinsky, Stephen R. Palumbi, James A. Estes, Francis H. Joyce, and Robert R. Warner, "Marine Defaunation: Animal Loss in the Global Ocean," *Science* 347, no. 6219 (2015), Article 1255641, doi:10.1126/science.1255641.

32. Andrew R. Solow and Andrew R. Beet, "On Uncertain Sightings and Inference about Extinction," *Conservation Biology* 28, no. 4 (2014): 1119–23.

33. Jake Kosek, *Homo-Apian: A Critical Natural History of the Modern Honeybee* (Durham, N.C.: Duke University Press, forthcoming).

34. For "planetary boundaries," see Will Steffen, Katherine Richardson, Johan Rockström, Sarah E. Cornell, Ingo Fetzer, Elena M. Bennett, Reinette Biggs et al., "Planetary Boundaries: Guiding Human Development on a Changing Planet," *Science* 347, no. 6223 (2015), Article 1259855. For "planetary stewardship," see Johan Rockström, Will Steffen, Kevin Noone, Åsa Persson, F. Stuart Chapin III, Eric F. Lambin, Timothy M. Lenton et al., "A Safe Operating Space for Humanity," *Nature* 461, no. 7263 (2009): 472–75, and Will Steffen, Åsa Persson, Lisa Deutsch, Jan Zalasiewicz, Mark Williams, Katherine Richardson, Carole Crumley et al., "The Anthropocene: From Global Change to Planetary Steward-ship," *Ambio* 40, no. 7 (2011): 739–61.

35. Timothy Lenton, Hermann Held, Elmar Kriegler, Jim W. Hall, Wolfgang Lucht, Stefan Rahmstorf, and Hans Joachim Schellnhuber, "Tipping Elements in the Earth's Climate System," *Proceedings of the National Academy of Sciences of the United States of America* 105, no. 6 (2008): 1786–93.

36. John von Neumann, "Can We Survive Technology?," *Fortune,* 1955, http://fortune.com/2013/01/13/can-we-survive-technology/.

37. Roger Fleming, *Fixing the Sky: The Checkered History of Weather and Climate Control* (New York: Columbia University Press, 2010).

38. Peter K. Haff, "Technology as a Geological Phenomenon: Implication for Human Well-Being," in *A Stratigraphical Basis for the Anthropocene,* ed. C. N. Waters, J. A. Zalasiewicz, M. Williams, M. Ellis, and A. M. Snelling (London: Geological Society Special Publications, 2014), 301–9.

39. von Neumann, "Can We Survive Technology?"

40. Anthony Costello, Mustafa Abbas, Adriana Allen, Sarah Ball, Sarah Bell, Richard Bellamy, Sharon Friel et al., "Managing the Health Effects of Climate Change," *The Lancet* 373, no. 9676 (2009): 1693–733.

41. Nixon, *Slow Violence.*

42. Michael Raupach and Joseph G. Canadell, "Carbon and the Anthropocene," *Environmental Sustainability* 2 (2010): 210–18.

43. Jackie Orr, "Slow Disaster at the Digital Edge," presentation to Worlding/Writing Project: In a Rut conference, University of Chicago, April 20, 2012.

44. Michelle Murphy, "Chemical Regimes of Living," *Environmental History* 13 (2008): 695–703. See also Kate Orff and Robert Misrach, *Petrochemical America* (New York: Aperture, 2012).

45. See also Nicole Six and Paul Petritsch, *The Sea of Tranquility* (Berlin: Landesgalerie Linz and Revolver, 2014).

46. Steffen et al., "Anthropocene."

47. Masco, "Catastrophe's Apocalypse," and Joseph Masco, "Bad Weather: On Planetary Crisis," *Social Studies of Science* 40, no. 1 (2010): 7–40.

5

Condors at the End of the World

Cary Wolfe

The question is indeed that of the world.
—Jacques Derrida, *The Beast and the Sovereign, Vol. 2*

What kind of event is extinction? To answer that question, we have to begin with an assertion that will seem paradoxical to some and commonsensical to others: that extinction is both the most natural thing in the world and, at the same time, is never and never could be natural. On one hand, 99.9 percent of all species that have ever existed in the history of this planet are extinct; on the other hand, extinction can hardly be regarded as "natural" in any simple sense, and not just because, as a number of people have argued, "nature," conceived as some realm apart, untouched and unshaped by human affairs, ceased to exist a long time ago, as all the recent talk about climate change and the Anthropocene makes clear.[1] Beyond this, the psychoanalytically inclined among us point out that any human registration of the so-called fact of nature is always already radically denaturalized because the symbolic and imaginary realms that register the presence of nature in their different ways for us are anything but "natural." As Slavoj Žižek put it, now nearly twenty-five years ago, "the fact that man is a speaking being means precisely that he is, so to speak, constitutively 'derailed,'" an "open wound of the world," as Hegel put it, that "excludes man forever from the circular movement of life," so that "all attempts to regain a new balance between man and nature" can only be a form of fetishistic disavowal.[2]

One way to intensify and complicate this assertion is to realize that everything Žižek says does not apply *only* to human beings (because we aren't the only "speaking beings" in the full sense intended here) even as we cannot say in any neat and simplistic way which nonhuman forms of life fall under the purview of this assertion and which do not—a desire for neat and tidy distinctions between different forms of life, different ways of being in the world, that would constitute its own form of fetishistic disavowal, as Jacques Derrida (among others) has pointed out. Indeed, as Derrida argues, in what may seem like a counterintuitive if not outlandish assertion,

> death is nothing less than an end of *the* world. Not *only one* end among others, the end of someone or something *in the world,* the end of a life or a living being. . . . Death marks each time, each time in defiance of arithmetic, the absolute end of the one and only world, of that which each opens as a one and only world, the end of the unique world, the end of the totality of what is or can be presented as the origin of the world for any unique living being, be it human or not.[3]

Now as we shall see in a moment, this seemingly brazen assertion can be redescribed in more naturalistic terms that make it seem a lot less counterintuitive. This will help us, in turn, draw out that what is going on here is not just an excessively Heideggerian hangover on Derrida's part; rather, he is trying to move us from what he calls the "dogma" of Heidegger's famous (or infamous) investigations of the differences between humans, animals, and stones to the inescapable necessity of paying attention to the different ways of being in the world, and it is on those differences that the hard and detailed ethical and political questions of thinking about extinction depend.[4] When a being, human or nonhuman, dies, what goes out of the world? When an entire species becomes extinct, what world leaves the world, the world we are left with? To begin to answer these questions is to realize that extinction, whatever else it may be, is never a generic event.

The California condor became extinct in the wild in 1987, when the remaining twenty-two individuals were captured and an ambitious conservation program was launched. In 1991, they were reintroduced into the wild, and as of October 2014, the total world population stands at 425 birds either in the wild or captivity, making it one of the world's rarest birds.[5] Bryndís Snæbjörnsdóttir and Mark Wilson's ambitious exhibition *Trout Fishing in America and Other stories* (2015) consists of several different elements, but the one I will focus on here is a series of photographs of the frozen, preserved bodies of fourteen dead California condors, each printed above a transcribed text about the bird taken from conversations with the biologists working in the conservation program.[6]

The trail that would lead to these photographs began when the artists visited the Vermilion Cliffs area of the Grand Canyon in November 2013 and interviewed biologist Chris Parish from the Peregrine Fund, who was working with the conservation program. During these conversations, the artists were especially struck by the contrast between the scientific protocols driving the project (having to do with body weight, medical regimes, and the like) and the often heartfelt and passionate stories they heard about each bird, mixing anger, frustration, and hope. It was here that they learned about the birds preserved in a freezer at the University of Arizona in Tucson—many of them dead owing to lead poisoning from feeding on animals killed by hunters with lead bullets. Later, once the photographs had been executed, they returned to Vermilion Cliffs and engaged in detailed discussions about each bird with Parish and fellow biologist Eddie Feltes, and it is from those conversations that the final printed transcriptions are taken.[7]

When I came to the opening of the exhibition, I was, at the same moment, in the middle of reading volume 2 of Jacques Derrida's seminars on *The Beast and the Sovereign,* in part in relation to another project I was working on (also about birds), and I felt a powerful resonance between Derrida's explorations of death, mourning, responsibility, and the concept of "world" (in the Heideggerian sense of the term) and the condor photographs—not just the images themselves but also the complex of emotions that bubble up between the lines (and sometimes not even between the lines) in the scientists' account of the life and death of each particular animal. To me, these images called forth the memorable line that ends Paul Celan's poem "Vast, Glowing Vault," which Derrida returns to again and again in the second volume: "The world is gone, I must carry you."[8]

Many things could be said, of course, about the relation between these lines and the condors that appear not only in the installation's photographs but also in its video components. For example, we might seize on the invitation offered by the figure of the "ram" in the poem, which evokes not only the primary reference of the constellation Aries but also, as Derrida notes, the ram in the story of Abraham and Isaac, the sacrificial substitute of animal for human which—paradigmatically, for Judeo-Western culture—secures the ontological privilege and specificity of the human "world" over and against that of the animal, to whom the biblical commandment "thou shalt not kill" evidently does not apply (*Beast 2,* 104–5). And yet, as Derrida points out elsewhere, "does killing necessarily mean putting to death? Isn't it also 'letting die'?"[9] I'll come back to complexities of that question at the very end of my chapter, but for now I want to pursue another invitation suggested by the "vault" of the poem's title and how it might be linked to the vault from which the condor bodies themselves have emerged to be placed before us in what strikes me, at least in some of these images, as a funereal setting or a scene of exhumation—an invitation to raise a question central to Derrida's seminars, and central to my response to these images, namely, what do we call these bodies before us? Are they "corpses," "remains"? Or just objects, like a rock, a table, or even a leaf? And if not, if they are remains, what are they remains *of*? To whom or to what do they belong? And what, in turn, is owed to them? Or as Derrida puts it in the dizzying opening of the fifth seminar in volume 2, this question of the corpse, which is "both a thing and something other than a thing" (*Beast 2,* 118), is a "stumbling block" but, at the same time, "an unavoidable touchstone" (*Beast 2,* 6) in our all-too-confident attempts to assume that human beings inhabit an ontologically full and secured world that is barred to animals, an assumption that makes the dead animal body not the remainder of something—or, better, some*one*—but rather nothing but an object on the order of a stone.[10]

To address these questions is perforce to ask, "What do beasts and men have in common?" (*Beast 2,* 8). As Derrida points out, Heidegger's own answer to this question is one that takes for granted a certain difference in kind between scientific knowledge, philosophical inquiry, and ethical and ontological questions—a difference that this particular art

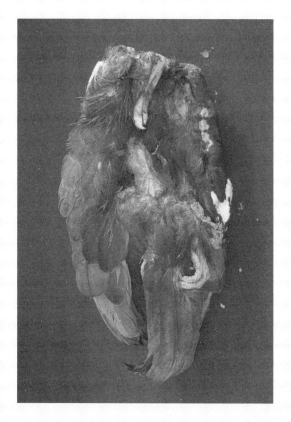

This bird was hatched on April 4th 1992. She was part of an experiment — her death was a coyote predation, so all that's left is a wing and a tail. She was an adult when she was released on the 7th of December 2000 but died December 25th the same year. She was only in the wild for a very short time and didn't have the necessary experience to deal with it. She wasn't conditioned — later on, when they had been in captivity for too long, we started doing things like conditioning them with dogs — training them with a dog. We hadn't done that before, so things like that were more likely to happen . . .

Figure 5.1. Bryndís Snæbjörnsdóttir and Mark Wilson. Condor 082, artwork from the exhibition *Trout Fishing in America and Other stories,* Arizona State University Art Museum, October 4, 2014, through January 17, 2015 (archival photograph, 2015). Courtesy of the artists.

installation seeks to trouble, and nowhere more so than in the texts that emerged from the conversations with the biologists whose dedication to their scientific protocols of observation and quantification could not conceal their intense emotional involvement with these creatures, these worlds, so different from our own.

No one limns this feature of Heidegger's approach to the difference between scientific and philosophical knowledge better than Derrida himself in the seventh seminar, when he notes that Heidegger's "strange concept" of the animal's "*poverty in world* . . . does not consist in a quantitative relation of degree, of more or less" (*Beast 2,* 192). "About this presupposed essence," he continues, "the zoologist, the zoologist *as such* at least, has nothing to say to us" (*Beast 2,* 194). As Michael Naas has pointed out, Derrida is often quick to note that Heidegger, more than most philosophers, takes "into account a certain ethological knowledge" with regard to animals.[11] But from Derrida's point of view, that only makes all the more dogmatic Heidegger's "thesis of essence ('the animal is poor in world') independent of zoological knowledge," a thesis pertaining, as Derrida puts it, to "the *animal in general,*" to "*every* animal" as "*equally* poor in world" (*Beast 2,* 196–97), whereas Derrida's focus will be not on "effacing the limit" between human and nonhuman animals (or indeed between different forms of life, human and nonhuman) but rather "in multiplying its figures, in complicating, thickening, delinearizing, folding, and dividing the line precisely by making it increase and multiply."[12]

Paying attention to the specificity and particularity of other forms of life is central to Derrida's fundamental strategy in laying out his differences with Heidegger in both *The Animal That Therefore I Am* and *The Beast and the Sovereign* seminars. As Naas characterizes it, Derrida's thought undertakes here a triple movement: first, he "begins by looking at a philosophical discourse that grants the human and denies the animal some attribute—language, technology, culture, mourning, a relationship to death, and so on"—and then second, he contests, "sometimes with reference to ethology, primatology, or zoology, the supposed *fact* that the animal does not have such and such an ability or attribute," while at the same time, and third, he "moves very quickly to the other side of the question in order to contest not the *fact* that animals do not have such and such a capacity or attribute but the *principle* on the basis of which philosophers have claimed that humans *do*" (26–27).

One such example—to return now, quite literally, to the question of "remains"—is Heidegger's well-known assertion that animals "perish," but only human beings "die," because human beings, unlike animals, have an understanding of death *as such*; they grasp their own mortality, and live in the light of it, in a way that eludes the animal, who simply ceases to exist biologically at the end of its life. And yet, as Derrida wonders in many places, do human beings really have this kind of relationship to death "as such," one that would allow this apparently radical form of finitude to be reappropriated as a "being-able," a "power" or a "potency" (*Beast 2,* 122)? Isn't it the case, rather, that we can never know the "as such" of death because death is always elsewhere and at a distance for us, even though it is, paradoxically, the thing that most testifies to our concrete and unique existence, our singularity?[13] After all, you can't experience your own death; you can only experience death in and through the death of the other, and all attempts to imagine or think about death are always, as Derrida points out, "phantasmatic." And "this suffices all the less," he continues, "to distinguish clearly between death as such and life as such because all our thoughts of death . . . are always, structurally, thoughts of survival. To see oneself or to think oneself dead is to see oneself surviving, present at one's death" (*Beast 2,* 117).

Here, then, we find not the finitude referenced by Heidegger—the confrontation with my mortality in his famous existential "being toward Death"—but rather what we might call *the finitude of my finitude,* its nonappropriability for and by me, its radical alterity, one that sets up a relationship of asymmetrical, unpredictable, and finally unappeasable alterity to the other. For as Derrida writes, "without knowing anything of what 'dead' means in the syntagma 'when I am, etc., dead,'" death means above all "to be delivered over, in what remains of me, as in all my remains, to be exposed or delivered over with no possible defense, once totally disarmed, to the other" (*Beast 2,* 126), and so "the other" names "what always might, one day, do something with me and my remains, make me into a thing," and do so, moreover, "as they wish" (*Beast 2,* 127), or as Derrida sometimes puts it, without "calculation."[14] To put it this way is to realize that this relationship of alterity and unpredictability to the other is, without letup and without assurances—indeed, *because* without assurances—a scene of ethical responsibility. And that is precisely the situation into which we are thrown, I would suggest, by these images. What shall we do with these remains that are "delivered over" to us? What will we make of them? And what will *that* make of us?

Here, I think, it's useful to augment Derrida's insistence on the alterity of other forms of life by redescribing it in terms of biological systems theory, but before we do, we need to follow the penultimate turn in Derrida's argument—to come back now to the last line of Celan's poem—which is made up of a movement through three possible theses, finished off with a meditation on the last:

> 1. Incontestably, animals and humans inhabit the same world, the same objective world even if they do not have the same experience of the objectivity of the object. 2. Incontestably, animals and humans do not inhabit the same world, for the human world will never be purely and simply identical to the world of animals. 3. In spite of this identity and this difference, neither animals of different species, nor humans of different cultures, nor any animal or human individual inhabit the same world as another, however close and similar these living individuals may be (be they humans or animals), and the difference between one world and another will remain always unbridgeable, because the community of the world is always constructed, simulated by a set of stabilizing apparatuses ... never natural, language in the broad sense, codes of traces being designed, among all living beings, to construct a unity of the world that is always de-constructible, nowhere and never given in nature. Between my world ... and any other world there is first the space and time of an infinite difference, an interruption that is incommensurable with all attempts to make a passage, a bridge, an isthmus, all attempts at communication, translation, trope, and transfer that the desire for a world ... will try to pose, impose, propose, stabilize. There is no world, there are only islands. (*Beast 2,* 8–9)

Now it is the first thesis that is usually taken to be "ecological," but my point here is that by the logic we are tracing, it is actually the *third* thesis that is the most radically ecological—or even better, we might say, *environmental.*[15]

Derrida's assertion might seem counterintuitive, but it will seem less so if we remember that in the terms of biological systems theory, "there is no world" precisely for the reasons we may trace back to the work of Jakob von Uexküll on human and animal *umwelten* and forward to those who work on the biology of consciousness and cognition, such as Humberto Maturana and Francisco Varela, who demonstrate that what counts as "nature" or "world" or (better still) "environment" is always a product of the contingent and selective practices deployed in the embodied en-action of a particular autopoietic living system. As philosopher of mind Alva Noë argues, "the locus of consciousness is the dynamic life of the whole, environmentally plugged-in person or animal."[16] And as his work shows, recent research in the biology of consciousness makes it clear that these questions do not neatly break along lines of human versus animal, inside versus outside, brain versus world, or even, for that matter, organic versus inorganic.[17] As Noë puts it, "it is not the case that all animals have a common external environment," because "to each different form of animal life there is a distinct, corresponding, ecological domain or habitat," which means, in short, that "all animals live in structured worlds" (43).

Now, I think, we are in a better position to follow Derrida as he moves rapidly in the next moment of the seminar from an equally bracing phrase taken from Daniel Defoe's *Robinson Crusoe*—the phrase "I am alone"—to Celan's memorable line that we have already quoted, "the world is gone, I must carry you." As Derrida puts it in a later session that year, picking up the thread,

> We could move for a long time, in thought and reading, between *Fort und Da, Da und Fort,* between . . . these two *theres,* between Heidegger and Celan, between on the one hand the *Da* of *Dasein* . . . and on the other hand Celan's *fort* in "Die Welt is fort" . . . the world has gone, in the absence or distance of the world, I must, I owe it to you, I owe it to myself to carry you, without world, without the foundation or grounding of anything in the world, without any foundational or fundamental mediation, one on one, like wearing mourning or bearing a child, basically where ethics begins. (*Beast 2,* 105)

Though I cannot pursue the point in detail here, it is worth noting that this scene of responsibility, generated by the absence of "world," is also a scene of what is sometimes called *spectrality* or *hauntology,* a scene of responsibility to those already gone ("wearing mourning") or those not yet here ("bearing a child"). Hauntology, as Colin Davis characterizes it, "supplants its near-homonym ontology" and marks "a wholly irrecuperable intrusion in our world, which is not comprehensible within our available intellectual frameworks, but whose otherness we are responsible for preserving." In this domain, the question of what the other asks or requires of us, a question secreted by its alterity, "is not a puzzle to be solved" but is rather "the structural openness or address directed toward the living by the voices of the past or the not yet formulated possibilities of the future."[18]

Now it is possible, in fact, to give a scientific account of how this hauntology or spectrality obtains in our relations to the living and to questions of extinction, an account that we are invited to pursue by remembering that paying attention to the nongeneric character of the system-environment relation for particular creatures is in fact quite consonant with Derrida's insistence on "complicating, thickening, delinearizing, folding, and dividing the line" that separates different forms of life—a move away from simplicity (namely, the simplicity of the ham-fisted distinction "human–animal") and toward *complexity* (namely, the nongeneric complexity of the system–environment relation as that evolves both ontogenetically and phylogenetically, a complexity that is, in Derrida's terms, infinite in principle). Indeed, as theoretical biologist and MacArthur Fellow Stuart Kauffman argues, the world is "enchanted" precisely *because* there are no "entailing laws" that govern, in a Newtonian fashion, the evolution of the biosphere and its various forms of life. As Kauffman puts it, even before we reach the level of what he calls "a Kantian whole, also sometimes called an 'autopoietic system' that 'builds itself'" (such as California condors),[19] we have to remember that

proteins are linear strings of amino acids bound together by peptide bonds. There are twenty types of amino acids in evolved biology. A typical protein is perhaps 300 amino acids long, and some are several thousand amino acids long.

Now, how many possible proteins are there with 200 amino acids? Well, there are 20 choices for each of the 200 positions, so 20^{200} or 10^{260} possible proteins with the length of 200 amino acids. . . . Now the universe is 13.7 billion years old and has about 10^{80} particles. The fastest time scale in the universe is the Planck time scale of 10^{-43}

seconds. If the universe were doing nothing but using all 10^{80} particles in parallel to make proteins the length of 200 amino acids, each in a single Planck moment, it would take 10^{39} repetitions of the history of the universe to make all the possible proteins the length of 200 amino acids just *once*! . . .

   History enters when the space of what is possible is vastly larger than what can actually happen. The universe in making a tiny fraction of all possible proteins the length of 200 amino acids is extremely *nonergodic*! (Kauffman 43)

Of course, as Kauffman points out, this principle obtains even more radically at the level of "Kantian wholes," since "the exploration of the ever vaster space of possibilities as the lengths of peptides and proteins increase . . . become ever sparser as complexity increases," which provides, in turn, "the basis of the antientropic process that undergirds part of why the universe became complex. This is surely true in the evolution of the biosphere," Kauffman continues, "from one or a few living things to hundreds of million species today with ever more myriad molecular and interwoven functional diversity of things and linked processes" (66). And from this vantage, what we confront in the bodies of these dead condors is, precisely, a materialized "trace," or even better, Derrida suggests, a "cinder" that he associates with a kind of "crypt" that prevents the completion of the work of mourning and psychic incorporation—one whose inscrutability haunts the present with retentions from an evolutionary past and protentions of an evolutionary future.[20] That is to say, as with any trace, this "cinder" confronts us with the presence of that which is not itself present—in this case, with the reality of the concatenative material processes that resulted in this form of life we call "condor" and with the evolutionary preadaptations present in the body before us that would, in time, take other, unpredictable forms, precisely, as Kauffman notes, because all possible organism–environment relationships in the future cannot be predicted or ergodically extrapolated from that which is present before us, try as we might to control or direct them.

   Nevertheless—and I've explored this question in some detail elsewhere—we do make decisions all the time, even in the face of this impossibility, without "foundational or fundamental mediation," about the "letting die" and the killing of various forms of life (human and nonhuman).[21]

And this leads to the final and far from trivial point about these animals foregrounded by this installation: that these bodies before us are part of an *archive,* one enmeshed in a complex landscape of legal, political, and scientific forms of knowledge and force, what Derrida calls those "stabilizing apparatuses" that simulate the sure and steady existence of a world in the face of the complexities we have just outlined. As Derrida points out, "there are no archives without political power," and, indeed, these condor bodies are in fact *evidence* of a potentially very charged political type, autopsied to reveal (more often than not) poisoning by a hunter's lead bullet. The archive is thus, as Derrida puts it, a kind of mise-en-scène of "two principles in one: the principle according to nature or history, *there* where things *commence*—physical, historical, or ontological principle— but also the principle according to the law, *there* where men and gods *command, there* where authority, social order are exercised, *in this place* from which *order* is given" (quoted in Naas 129). What better way to mark this fact, in these images, than the strange cohabitation, within the same frame, the same "place," of these singular dead animal bodies, subject to the laws of chemistry, decay, rigor mortis, and the like—"ultra-natural" objects, in that sense, whose decay we try to control through technological means—and what Derrida has called the *machinalité* of any semiotic code whose epitome is, of course, mathematics, here represented by the "anonymous" numbers that mark each bird's wing tag but only to become, in time, a kind of emotionally charged "proper name" for *this* particular creature—all of which redoubles and accumulates in the seriality of the photographic series itself.

What this means is that as it is with the archive, so it is with extinction. On one hand, there is nothing more "natural" than extinction—it is an event that happens "*there,*" in nature, and *has* happened with the vast majority of species that have ever existed; but at the same time, extinction is and can never be a "natural" event because it always takes place within an horizon of "world" and its governing principles—including, of course, the principle of "biodiversity"—that *we* create through "stabilizing apparatuses." But that stabilization is always marked by something else that is preserved—maybe even *mainly* preserved—in the archive: what Derrida calls the *adestination* or *destinerrance* that attends any attempt

A female hatched on the 23rd of May 1999 in Boise, Idaho. We released her back on December 28th the same year. She was in the wild until 1st of February 2013 when she died of lead poisoning. From the beginning of her time in the wild she was a real problem. She had absolutely no fear of people and of course our concern with that was that people might feed her. She turned out to be one of the best birds of the population and she also produced young in the wild. So that's a tough thing, losing an adult, producing bird. A hard hit to the population . . . We thought if we would have been able to track her we would have taken her and put her in captivity forever. She was a bird that had absolutely no fear. Generally I would say that Condors don't have much fear of humans and that lack is one of our biggest problems. But we were ready to catch her — she would land at the jewelry stands on the side of the highway and she'd land near parked cars, probably because she had been fed; they're certainly not stupid. Then she turns out to be one of our best birds.

Figure 5.2. Bryndís Snæbjörnsdóttir and Mark Wilson. Condor 210, artwork from the exhibition *Trout Fishing in America and Other stories,* Arizona State University Art Museum, October 4, 2014, through January 17, 2015 (archival photograph, 2015). Courtesy of the artists.

to make good on our commitments, to materialize our "world," to address the other to whom we feel responsible: an *adestination* that stems
from the fact that the same sign or trace or mark can function variably,
even oppositely, in very different contexts.[22] The constraints of scientific
method and protocol constitute, of course, a canonical attempt to control,
even eliminate, this *destinerrance,* but its most compelling manifestation
in this installation is the lead bullet that leaves its trace, sometimes in
the discoloration of the animal's body by lead poisoning, but sometimes
invisible, in these bodies and these images but not *of* them, you might
say, the materialization of two "*there*s" in one "place." That *destinerrance*
quite literally attends such tidy ethical, legal, and political distinctions
as we like to make between the polar opposites of "game" or "trash"
animals who are deemed "killable but not murderable"—the animals
that sustain these carrion feeders—and those who, like the condor, are
"rare," "threatened," and "protected," with the full backing of scientific
and political apparatuses. The archive, in other words, may record the
"official story" of body weight, reproductive rate, legal status, and so on,
but it also actualizes something more, and in that other space, that other
scene, we discover that the world is not given but made. We thus discover,
in short, a scene of responsibility.

NOTES

1. For one version of this argument, see Timothy Morton, *Hyperobjects: Philosophy and Ecology after the End of the World* (Minneapolis: University of Minnesota Press, 2013).

2. Slavoj Žižek, *Looking Awry: An Introduction to Jacques Lacan through Popular Culture* (Cambridge, Mass.: MIT Press, 1992), 36–37.

3. Jacques Derrida, "Rams: Uninterrupted Dialogue—between Two Infinities, the Poem," in *Sovereignties in Question: The Poetics of Paul Celan,* ed. Thomas Dutoit and Outi Pasanen (New York: Fordham University Press, 2005), 140.

4. Derrida's characterization of Heidegger's "dogma" appears in "*Geschlecht* II: Heidegger's Hand," trans. John P. Leavey Jr., in *Deconstruction and Philosophy,* ed. John Sallis (Chicago: University of Chicago Press, 1986), 173.

5. Wikipedia, s.v. "California condor," last modified December 16, 2016, http://en.wikipedia.org/wiki/California_condor.

6. Bryndís Snæbjörnsdóttir and Mark Wilson, *Trout Fishing in America and Other stories,* Arizona State University Art Museum, October 4, 2014–January 17, 2015. See http://snaebjornsdottirwilson.com/category/projects/trout-fishing-in-america/.

7. Details about the process that eventuated in the photographs were reported to me in an e-mail from Mark Wilson on December 30, 2014.

8. The translation I use here is from *Poems of Paul Celan,* trans. Michael Hamburger (New York: Persea, 2002), 275. Derrida first cites the passage in *The Beast and the Sovereign, Volume 2,* ed. Michel Lisse, Marie-Louise Mallet, and Ginette Michaud, trans. Geoffrey Bennington (Chicago: University of Chicago Press, 2011), 9. Further references to Derrida's text are given internally as *Beast 2*.

9. Jacques Derrida, interview with Giovanna Borradori, *Philosophy in a Time of Terror: Dialogues with Jürgen Habermas and Jacques Derrida,* ed. Giovanna Borradori (Chicago: University of Chicago Press, 2003), 108.

10. Derrida very much intends the play between "stone" (drawn from Heidegger's canonical discussion of the ontological differences between stones, animals, and humans) and "touchstone" (*Beast 2,* 6). For two exemplary literary instances that worry this question in especially powerful ways, see the latter chapters of J. M. Coetzee's novel *Disgrace* (New York: Penguin, 2000) and the behavior of its central character, David Lurie, toward the dead bodies of dogs killed at the animal shelter at which he works, and, in nonfiction literature, Berry Lopez's *Apologia* (Athens: University of Georgia Press, 1998).

11. Derrida, quoted in Michael Naas, *The End of the World and Other Teachable Moments: Jacques Derrida's Final Seminar* (New York: Fordham University Press, 2015), 149. Further references to Naas's book are given in the text.

12. Jacques Derrida, *The Animal That Therefore I Am,* ed. Marie-Louise Mallet, trans. David Wills (New York: Fordham University Press, 2008), 28. See also 34–35.

13. See, e.g., David Farrell Krell's discussion in *Derrida and Our Animal Others: Derrida's Final Seminar, "The Beast and the Sovereign"* (Bloomington: Indiana University Press, 2013), 66.

14. See, in this connection, Derrida's interview "'Eating Well' or the Calculation of the Subject," in *Who Comes After the Subject?*, ed. Eduardo Cadava, Peter Connor, and Jean-Luc Nancy, 96–119 (New York: Routledge, 1991).

15. My colleague Timothy Morton has made his own version of this argument—that ecological thinking *begins* with what he calls "the end of the world." See, e.g., Morton, *The Ecological Thought* (Cambridge, Mass.: Harvard University Press, 2012), and Morton, *Hyperobjects*.

16. Jakob von Uexküll, *"A Foray into the Worlds of Animals and Humans," with "A Theory of Meaning,"* trans. Joseph D. O'Neil, with an introduction by Dorion Sagan (Minneapolis: University of Minnesota Press, 2010); Humberto Maturana and Francisco Varela, *The Tree of Knowledge: The Biological Roots of Human Understanding,* rev. ed., trans. Robert Paolucci, with a foreword by J. Z. Young (Boston: Shambhala Press, 1998); Alva Noë, *Out of Our Heads: Why You Are Not Your Brain, and Other Lessons from the Biology of Consciousness* (New York: Hill and Wang, 2009), xiii. Further references to Noë are given in the text.

17. For more on how these questions cross-pollinate with Derrida's work, see my book *Animal Rites: American Culture, the Discourse of Species, and Posthumanist Theory* (Chicago: University of Chicago Press, 1998), 78–94, and, more recently, *Before the Law: Humans and Other Animals in a Biopolitical Frame* (Chicago: University of Chicago Press, 2013), 60–86.

18. Colin Davis, "*État Présent*: Hauntology, Spectres and Phantoms," *French Studies* 40, no. 3 (2005): 373, 378–79.

19. Stuart Kauffman, *Humanity in a Creative Universe* (New York: Oxford University Press, 2016), 67. Further references are given in the text.

20. See Jacques Derrida, *Cinders,* trans. Ned Lukacher, with an introduction by Cary Wolfe (Minneapolis: University of Minnesota Press, 2014).

21. See the last two chapters of Wolfe, *Before the Law.*

22. See J. Hillis Miller, "Derrida's *Destinerrance,*" *MLN* 121, no. 4 (2006): 896.

6

It's Not the Anthropocene, It's the White Supremacy Scene; or, The Geological Color Line

Nicholas Mirzoeff

This essay is by way of a provocation and an opening to a broader discussion. It is the result of asking, What does it mean to say #BlackLivesMatter in the context of the Anthropocene? As is now common knowledge, the Anthropocene is the proposed name for a new geological era, the "recent human era." This understanding relies on the identification of a "single physical manifestation of a change recorded in a stratigraphic section, often reflecting a global-change phenomenon."[1] This ability to perceive and agree upon a visible and graphic distinction in physical phenomena is inevitably and persistently imbricated with concepts of race and racialization from the very formation of what is now called Earth system science.[2] In short, my question is, What kind of "man" is meant when we say *Anthropocene*? Given that the Anthropos in *Anthropocene* turns out to be our old friend the (imperialist) white male, my mantra has become, it's not the Anthropocene, it's the white supremacy scene. Many within academia might find such terminology too crude or extreme. For #Black-LivesMatter activists in the United States (and now Britain), however, white supremacy is a given. Since the events in Ferguson following the shooting of unarmed teenager Michael Brown by then officer Darren Wilson, even mainstream figures like Hillary Clinton have been speaking of "systemic racism," a term previously used by activists and academics.[3]

In the time frame of the Anthropocene (whichever one uses), that system can only mean "white" (Euro-American) domination of the colonized and enslaved African, Asian, and Native populations of the world.

In this context, moreover, the term *life* is key. The concept of extinction itself was part of the transformation of natural history into life science (biology) in the era of the revolutions of the enslaved and abolition (1791–1863). In the current paradigm shift to Earth systems (everything prefaced with *geo-*), what is the place of Black and other colonized life, human and nonhuman? Has the political failure to enact change in relation to the crisis of the Earth system not been motivated precisely by systemic racism? This racism has numerous dimensions: the vastly higher CO_2 emissions per capita in the United States than any other nation (with the exception of small, oil-producing countries); the pollution hot spots in people-of-color communities within the United States; the ecological disaster of mass incarceration that has been called prison ecology, and so on.[4] This constellation has not escaped scholarly attention at all.[5] I want to add to this discussion by proposing that the very concept of observable breaks between geological eras in general and the definition of the Anthropocene in particular is inextricably intermingled with the belief in distinct races of humanity in the former instance and the practices of (neo)colonialism in the latter, centered on questions of the definition of life, how to make distinctions, and how to see difference. Any Anthropocene politics would, in turn, need to begin by being antiracist and anticolonialist.

In what follows, I concentrate on the two moments of definition: the formation of the concept of extinction and geological eras in the age of abolition and the revolutions of the enslaved, and the ongoing and contested debate over the Anthropocene. In the first section, I show how geology and race theory combined to produce a color line enshrined in natural history rather than law in the wake of the abolition of slavery. This emphasis on "line" could be found across Euro-American culture from the natural sciences to painting. No sooner had the concept of extinction been announced by Georges Cuvier in the early nineteenth century than its founder was hard at work trying to define an essential and visible "line" or difference between Africans and Europeans in the wake of the revolution of the enslaved and the abolition movements. In the second section,

I engage with the spectacular and public ongoing debate in present-day geological stratification as to whether the Anthropocene was the result of intentional, if misguided, world shaping by Euro-Americans or the consequence of colonial and imperial ambition. The former now seems to have the bureaucratic advantage and, as it were, sets in stone the history of white supremacy as geology. In so doing, the Anthropocene has become a measure of human time rather than a marker of physical processes. Humans are now claimed to be geological masters, a term that should give us pause in the context of slavery and racism.[6]

There is now a substantial body of humanities scholarship being produced in response to the combined impact of the Anthropocene turn, the material turn, and the nonhuman turn. It is an inspiring and important development. We should nonetheless recognize that the cumulative effect has been to generate a turn away from understandings of race, white supremacy, colonialism, and imperialism, which undermines the possibility of a politics of resource use and allocation, also known as the commons.[7] In the introduction to their important 2003 volume on race and nature, Donald Moore, Jake Kosek, and Anand Pandian argued, "Race provides a critical medium through which ideas of nature operate, even as racialized forces rework the ground of nature itself."[8] One of the key themes of Anthropocene writing has been the idea that nature has been replaced or overdetermined by human activity. In that overwriting, the central function of race within the framing of the Earth system has been displaced. This discursive move is not intentionally racist, except insofar as it is a mark of a certain privilege to be able to overlook race. My anxiety with the material, nonhuman, and universalist turns in academic discourse is, then, how quickly we seem to forget all the work that has been done to establish how and why so many people have been designated as nonhuman and bought and sold as material objects. Take a canonical example: the 1857 Supreme Court decision *Dred Scott v. Sandford* that enshrined the legal distinction between "the dominant race" and the "subordinate and inferior race of beings" known as the "negro African race," to use the terms of Chief Justice Taney's ruling opinion.[9] For Taney, "a perpetual and impassable barrier" existed in the laws of the thirteen colonies between the two groups, meaning that the Declaration of Independence's assertion of liberty for all could not apply to those who were simply an "article of property." Enslaved people were always and already not part of any universal. They were, however, objects (in the eyes of the enslavers). This "barrier," line, or break was a palpable (if later discredited and displaced) part of framing life sciences such as geology.

This awareness of the enslaved as nonhuman objects is not visibly a part of those whose work is indispensable for everyone now thinking about the Anthropocene. In Jane Bennett's instant classic of the nonhuman turn, *Vibrant Matter,* for example, there is no discussion of race in relation to vital materialism. Race is, of course, still present, because it cannot not be. It arises symptomatically. The denial of agency to matter central to Bennett's agenda (and with which I am sympathetic in itself) is equated via a quotation from Bruno Latour to the moment "when the Founding Fathers denied slaves and women the right to vote."[10] Quite apart from the false equation of slavery and the right to vote, what happened to understanding the chattel part of chattel slavery, so central to *Dred Scott*? An enslaved person was an "article of property," an object, nonhuman and commodified. Whether we agree with this classification or not, we must accept its immense and continuing significance. Race, in Bennett's account, is a problem only for what she calls political ecology, not the theorizing of materialism.[11] The costs of sidelining this politics are clear in her book, when she endorses Garret Hardin's statement in his "Tragedy of the Commons" essay that "freedom in a commons brings ruin to all."[12] While we should note his neoliberal misrepresentation of common stewardship of the land, let us concentrate here on whom Hardin blamed for the population crisis he envisaged as "tragedy." Writing in 1968, of all years, he disparaged those whose "cries of 'rights' and 'freedom' fill the air." Hardin's fear of the commons was more exactly fear of a Black planet.

This is not just an issue of priorities. For the historian Dipesh Chakrabarty, among the first humanists to use the term, "in the era of the Anthropocene, we need the Enlightenment (that is, reason) even more than in the past."[13] This reason identifies humans as a species, an identification that we cannot experience but only intellectually comprehend or infer. Thus he calls for a "new universal history" to comprehend the species. It is often said that climate change affects everyone, but it does not do so equally. During Hurricane Sandy in October 2012, there was undoubtedly tragic loss of sixty-seven people in the United States, who died as a direct result of the storm, and a further thirty-eight indirect deaths.[14] By contrast, Typhoon Haiyan in November 2013 caused what can only be estimated as between six and ten thousand deaths in the Philippines. That is to say, while climate change certainly affects the entire planet, its impact is very different in different places, consistent with the usual indicators of wealth. Furthermore, as Jacques Rancière has put it, "democracy can never be identified with the simple domination of the universal."[15] Indeed, the formation of the idea of "one world" can be seen as the connection of Christianity's belief in the human dominion of nature (including non-human humans) with capitalism.

So we should ask how such universal history might be written. In 1938, C. L. R. James declared that the enslaved field workers in Saint-Domingue (later Haiti) were the first modern proletariat.[16] In 1944, the historian and politician Eric Williams connected capitalism and slavery in his book of that title.[17] It has taken more than half a century for mainstream (white) history to come around to this point of view, based on studies of the vital place of cotton in developing the U.S. economy, which provided "a rocket booster to American economic growth."[18] In an additional twist, former enslavers in Britain used the compensation they received when slavery was abolished to launch themselves as capitalists. British prime minister William Gladstone's family benefited to the equivalent of £3 million in present-day money, while the enslaver Nathaniel Snell Chauncy directed in his 1848 will that all his property in the Caribbean should be sold and invested in railways.[19] If universal history is the history of how capitalism has produced globalization, that history in turn is also the history of enslavement, which, together with the transatlantic exchange of plants, goods, people, animals, and viruses caused by the triangular trade, is newly central to understanding the Anthropocene, as we shall see. Cash crop cultivation by enslaved human beings in the Caribbean and Indian Ocean was itself responsible for some of the first systemic anthropogenic environmental catastrophes on islands such as Barbados, Jamaica, and Réunion.[20]

Furthermore, as the Brazilian scholar Denise Ferreira da Silva has shown, the Enlightenment thought that creates the concept of universal history itself depends on what she calls "the core statement of racial subjection: while the tools of universal reason (the 'laws of nature') produce and regulate human conditions, in each global region it establishes mentally (morally and intellectually) distinct kinds of human beings, namely, the self-determined subject and its outer-determined others, the ones whose minds are subjected to their natural (in the scientific sense) conditions."[21] That is to say, in the Enlightenment concept of universal reason, certain people are produced by "nature" as those worthy of being colonized or open to being enslaved. There is, then, no "innocent" nature that was later despoiled by the Anthropocene: the very idea of nature is inextricably entangled with race. As da Silva argues, the Enlightenment project is to create both a "scene of regulation, which introduces universality as the juridical descriptor," and a "scene of representation" in which only the colonizer has the interior judgment capable of recognizing and interpreting representation. As Marx famously ventriloquized this view, "they cannot represent themselves, they must be represented."[22] Such representation engages both meanings of the term: political representation and visual or cultural depiction are interfaced aspects of the same violent relationship, securing "the stage of interiority" to render judgment on what is exterior.[23]

THE GEOLOGICAL COLOR LINE

This scene of representation has a clear counterpart in modern geology, which has stressed the need to find visible points of transition from one geological epoch to the next. The opening words of Charles Lyell's classic *Principles of Geology* (1830) make this clear: "Geology is the science which investigates the successive changes that have taken place in the organic and inorganic kingdoms of nature."[24] Change was visible in "lines of demarcation" between geological strata. As such, it embraced all the "physical sciences," including zoology, as well as classificatory theories like "natural philosophy." Modern geology has perhaps moved away from such all-embracing views, but its creation of the geologic time scale has been hailed as one of the great achievements of science, just as Lyell imagined. At the same time and in the same historical moment, meaning the early nineteenth century (1791–1848), the tendency to stress a hypervisible point of distinction was formed in the racist effort to define extinction after the abolition of slavery as part of a system that sustained the concept of separate species of human being. Stratigraphy was shaped by the doubled desire to mark the historic eras of Earth's history *and* to trace a systemic boundary between races as a means of containing and displacing abolition and revolution.[25] This break was conceptual but visible—at least to its protagonists, who insisted on the special and refined form of visual observation required to perceive it. The line they saw was one derived from neoclassical painting and enshrined in photography that continues to structure our observation today. Here are a cluster of examples from key figures in modern science who both made remarkable breakthroughs in understanding that shaped the interface of geology with climate and natural history and sought to insist on racial distinction.

Georges Cuvier

Georges Cuvier, who is usually credited with defining extinction, described it as a form of revolution in 1807, immediately following the independence of Haiti in 1801. His concept was part of his reconsideration of natural history as the science of life. This turn to life was, as Michel Foucault long ago noted (and an army of followers have since), foundational for modern sciences and societies. Cuvier elaborated on his study of the "conditions of existence" in his 1802 study *The Animal Kingdom*. Having outlined his theory of extinction, Cuvier at once developed his race theory.[26] He divided humans into three varieties of which "the Negro race is confined to the south of Mount Atlas. It is marked by a black complexion, crisped or wooly hair, compressed cranium, and a flat nose. The projection of the lower parts of the face, and the thick lips, evidently approximate it to the monkey tribe; the hordes of which it consists have always remained in the most complete state of utter barbarism."[27] In the very next sentence after his assertion of African barbarism, he continues, "The race from which we are descended has been called *Caucasian*," the euphemism still used for *white*. There was, then, a doubled line to be seen. It marked both extinction and barbarity. Often the deduced barbarity would be held to justify separation and slavery. Its logic implied that extinction would result, whether by human or "natural" process, leading Cuvier to notoriously define native Africans as "the most degraded of human races, whose form approaches that of the beast and whose intelligence is nowhere great enough to arrive at regular government."[28] This despite the fact that a majority African population had just done that in Haiti, a former French colony, of which Cuvier could not possibly be ignorant. For Cuvier, then, one cannot even make the phrase "Black lives matter" make sense. Black life for him was a variety of animal life, whose outcomes might be the subject of curiosity but not moral engagement. It was this same Cuvier who engaged in the dissection of the deceased Sara Baartman, a Khoisan woman (from present-day South Africa), and preserved her genitalia for the collections of the Museum of Natural History. Cuvier argued that geological as well as human history was marked by a series of catastrophes. The "line" was both the mark of past (geological) catastrophe and the means of defense against future (human) recurrence. The line is not single or definitive but a place of ambivalence and anxiety. Welcome to the Anthropocene.

Jean-Jacques Audubon

For the slave owner turned naturalist Jean-Jacques Audubon (also known as John James Audubon, Jean-Jacques La Forêt, and John James Forest), these interconnections were very personal.[29] Born in Saint-Domingue to an enslaver father and a Jewish servant mother, Jeanne Rabin, he became a refugee from post-independence Haiti, haunted by abolition and the extinction of birds and of the Native population—whom he saw as doomed—and indeed of the American wilderness. In this way, his ornithology, as a subset of zoology, mapped change in a fashion that interfaced with Lyell's contemporary definition of geology. Audubon turned to writing about birds after his debt-funded purchase of slaves to work at his Kentucky mill ended in bankruptcy in 1819, writing in his 1826 Mississippi River journal that "so *Strong* is my Anthusiast to Enlarge the Ornithological Knowledge of My Country that I felt as if I wish myself *Rich again.*"[30] His last human property rowed him down the Mississippi to New Orleans, where he sold the two men.

His famous drawings of birds were less original than most of us think.

Their large format and "action" poses were in fact standard at the time in French (if not North American) ornithology.[31] Audubon's scenes were not taken from life but drawn using dead birds suspended by wires, a technique he claimed to have derived from the studio of the great neoclassical painter Jacques-Louis David. While Audubon did have artistic training in France, there is no record of his having been part of David's extensive studio. Perhaps he learned from someone who had been there, or, more likely, he made it up. David's style certainly centered on the depiction of line to create form, just as Audubon's work does. In painting, to use "line" was to define your work as History, meaning its most serious and morally important category, used to depict momentous events in biblical and human history. Perhaps the only ornithological artist to have aspired to—and achieved—that status was Audubon. And by claiming the lineage of the great artist, he also displayed the instinct of the showman that he certainly was. His originality was to express the tensions between race, colonization, and extinction in a nonhuman but evocative form of History, which is to say, birds. When he looked at birds, for all his noted powers of observation, other phantasms filled his vision.

Late in his life, Audubon sought to explain why he had become a bird artist in his autobiographical sketch, *Myself*:

> My mother had several beautiful parrots and some monkeys; one of the latter was a full-grown male of a very large species. One morning, while the servants were engaged in arranging the room I was in, "Pretty Polly" asking for her breakfast as usual, . . . the man of the woods probably thought the bird presuming upon his rights in the scale of nature; be this as it may, he certainly showed his supremacy in strength over the denizen of the air, for, walking deliberately and uprightly toward the poor bird, he at once killed it, with unnatural composure. The sensations of my infant heart at this cruel sight were agony to me. I prayed the servant to beat the monkey, but he, who for some reason preferred the monkey to the parrot, refused. I uttered long and piercing cries, my mother rushed into the room, I was tranquillized, the monkey was forever afterward chained.[32]

As a primal scene of the white supremacist imagination, this can scarcely be bettered. It centers on the doubled figure of "the man of the woods" *(homme de la forêt),* that is to say, on the orangutan and Audubon himself, who also used the name La Forêt. Then again, young Audubon is also in the scene directly, witnessing the death of the parrot Migonne, who, as Christopher Ianinni observes, is a symbol for either his birth mother or his adopted mother. It can also be seen as a transposition of the enslavers' view of the Haitian revolution into the not quite human world. The ape, in this view, kills "the feminized representative of European refinement."[33] The French-speaking parrot loses its life to the simian claiming his rights. Neither of the Audubons quite gets what he wants. The orangutan is not whipped—and do we not see here a desire for the whipping of an enslaved person, transposed (not very far in the racist imagination) from Africans to apes? If La Forêt is the "monkey," he loses by being clapped in chains, like a punished enslaved person. In this unresolved set of sexualized and racialized fantasies, it is "the scale of nature" that becomes the measure of the different standing of the different persons. The point is precisely the failure of such a scale to measure up, letting down all the different threads.[34]

It might be said that this reading goes too far. Yet it was a repeated figure in Audubon's work. In his *Ornithological Biography,* published as a textual accompaniment to the famous pictures, he had already concocted a similarly bizarre fantasy.[35] Lost in the Louisiana woods, Audubon claims to have encountered a maroon (whom he calls a runaway slave) living in a cane brake with his family.[36] Here is yet another *homme de la forêt,* or as Audubon put it, mixing racialized metaphors, "a perfect Indian in his knowledge of the woods." Developing his story, Audubon tells how this man had been resold following the bankruptcy of his first owner, separating his family. He memorized the destination of his wife and children and, after he himself had escaped, rescued them and, with the cooperation of those still enslaved, made camp in the woods. The bankruptcy and family breakup again echo Audubon's personal, rather than ornithological, biography. He devised a fantasy ending in which the maroons obeyed him because of their "long habit of submission" and returned with him to their original plantation, where Audubon persuaded the new owner to take them all into his ownership. He ends his little reverie with the inaccurate statement that since this time it has "become illegal to separate slave families without their consent." The pursuit and biography of birds led Audubon to imagine personal and political reconciliation within racial hierarchy and restored slavery, as if the Haitian revolution had never happened. The History he paints in these words is, one might say, nonlinear.

Louis Agassiz

On the other side of the Atlantic world, another Francophone naturalist was defining a different form of extinction by means of precise observation. The Swiss naturalist Louis Agassiz studied the glaciers in the Alps and came to define the ice ages. His work was always interfaced with a curious obsession with creation and extinction that became specific around race. Like the other protagonists in this survey, Agassiz claimed to have what one might call hypervision. He recommended a special diet for naturalists "in order that even the beating of his arteries may not disturb the steadiness of his gaze, and the condition of his nervous system be so calm that his whole figure will remain for hours in rigid obedience to his fixed and concentrated gaze."[37] Long before the "male gaze" had been named, Agassiz invented a disembodied naturalist gaze that becomes the viewpoint of God. The key was the observation of striations, faint scratches on rocks that could not have been produced by water, which tends to polish, but indicated the result of past friction. By observing such striations on rocks, polished surfaces, and unusual locations of so-called erratic boulders shaped, smoothed, and moved by glacial action, Agassiz was able to deduce the past presence of the ice ages. As he noted, had the boulders been left there by a flood (or the Flood), they would have been polished rather than striated. For all his claims to insight, the extended glacier theory was first suggested by a peasant named Perraudin, who told the idea to a professor named Jean de Charpentier, who in turn relayed it to Agassiz.[38] Agassiz realized that these expanded glaciers were the cause of past extinctions. He came to theorize extinction (*disparition*) as a repeated event: "There is therefore a complete scission between current creation and those which have preceded it." Resemblances between living species and extinct ones is a matter of "progeniture," or they are simply "identical species" re-created.[39] Geological time is in this view, like that of Cuvier, divided by radical breaks that nonetheless were followed by simulacra of what had gone before.

Glaciers, he noted, cannot simply be seen by looking.[40] Their history is deduced like faint traces on a lithographic stone, a literally fossilized photograph, creating an intriguing co-incidence with the concurrent attempt to create photography, realized by both Daguerre and Talbot in 1839.[41] Agassiz proposed that these epochs of creation and disappearance could be represented as a pattern in which a sudden catastrophic drop in temperature was followed by a later warming, although not to the same high temperature as before. At each upturn, there was a new moment of creation. Agassiz drew a curious pseudo-graph to illustrate this dynamic, consisting of straight lines connected by the reversed letter *J* to convey this sense of catastrophic transformation. These revolutions were visible only as "a prerogative of the scientific observer, who can tie together in their mind facts which appear without connection to the mob."[42] For any nineteenth-century European bourgeois reader, the mob was itself always connected to revolution. The revolution was, by extension, blind, lashing out wildly in what Thomas Carlyle called in his contemporary *History of the French Revolution* "chaos," in which "dim masses, and specks of even deepest black, work in that white-hot glow of the French mind, now wholly in fusion and *con*fusion."[43] Bad as this was, it paled next to Haiti: "Black without remedy."[44] Parenthetically, Agassiz's concept of irregular evolution was not entirely as eccentric as it might seem. One of the most widely discussed ideas in recent studies of evolution has been that of "punctuated equilibrium," proposed by Niles Eldredge and Stephen Jay Gould.[45] On the basis of observation of the paleontological record, Eldredge and Gould argued for periods of dramatic change (over the course of fifty to one hundred thousand years) followed by extended stasis. Gould, of course, was a legendary debunker of scientific racism, whose classic *Mismeasure of Man* included a pioneering documentation of Agassiz's racism.[46]

Like Cuvier, Agassiz soon turned his observational gaze toward race. As is well known, having moved to the United States to be a Harvard professor, he commissioned the daguerreotypist J. T. Zealy to take a series of shots of enslaved Africans in South Carolina in 1850. Rediscovered in the Harvard library in 1973, these pictures are among the first known photographs of the enslaved. Agassiz expected to see in these bodies incontrovertible visual evidence of racial difference, which is to say, the color line. His goal was that a sophisticated (white) person would be able "to distinguish their nations, without being told whence they came and even when they attempted to deceive him," just as he had detected the presence of the ice ages and catastrophic climate change.[47] Agassiz demonstrated a poor understanding of the practice of slavery, even as he was up to date in race "science." In the slave market, one might observe, as Joseph Holt Ingraham put it in 1835, "that singular look, peculiar to a buyer of slaves."[48] This look was less concerned with national origin, following the formal abolition of the slave trade, than with the degree of whiteness in the chattel body. Blackness was preferred for outdoor work, whereas lighter-skinned women were picked for household labor, but so too was it important where in the United States an enslaved person had grown up. Close observation of the hands, arms, teeth, and face was considered essential. In the parallel vision of race science, as Molly Rogers has pointed out, the claim to see a person's origin at a glance was common. The naturalist Samuel Morton also claimed to able to deduce ethnicity from skulls with "a single glance of his rapid eye."[49]

Zealy's photographs for Agassiz were forgotten precisely because the project of making national—meaning racial—origin visible failed. When they reemerged, photographic historian Alan Trachtenberg hailed their "universal humanness," looping the desire for visible distinction back into the (white/colonial) universal.[50] Others have followed Brian Wallis in seeing the enslaved refuse all cooperation with the measuring project.[51] Zealy posed his figures as if for the slave market, stripped to the waist. Indeed, photographs were used to advertise human beings for sale. Although Agassiz professed to oppose slavery, he argued for separate and distinct "races" of human beings, known as polygenesis, as a corollary to his theory of repeated creation and extinction. Such theories were a staple reinforcement of slavery—because enslaving a different species is different to enslaving what abolitionists called "a man and a brother"— and were decisively overturned by Darwin in *On the Origin of Species* (1859). Agassiz's deduction of the ice ages from close visible observation of linear marks is part of commonsense knowledge today, but we cannot neatly separate it from his conviction that a color line could be seen on the human body that he maintained in the face of the theory of evolution.

Richard Owen

In 1839, the traditional founding date for photography, British naturalist Richard Owen identified the moa as an extinct flightless bird from one bone found in New Zealand. He read the lines within the fossil to correctly demonstrate that it was a flightless bird, not a large animal. To do so, he created a form of verbal hyperdescription that is both analogous to photography and connects such natural history to the classic realist novel of the period: "The exterior surface of the bone is not perfectly smooth, but is sculptured with very shallow reticulate indentations; it also presents several intermuscular ridges. One of these extends down the middle of the anterior surface of the shaft to about one-third from the lower end, where it bifurcates."[52] From this beginning, Owen went on to deduce the existence of more than eighty species of moa, where scientists now think there were perhaps eleven. Unwilling to accept theories of evolution, like Agassiz, Owen argued that the moa was in fact a degeneration from earlier winged species, following the theories of Buffon and Lamarck. In support of this idea, designed to be a rebuttal to Darwinism via the moa, Owen produced an extraordinary volume of hyperreal, almost three-dimensional lithographs of moa bones. More photographic than photographs, the repetitive drawings extend unevenly out of the covers of the book to a length sometimes reaching two or three feet. Mostly produced by James Erxleben (circa 1830–80), these works are masterpieces of a certain kind, an astonishing precision and detail in the service of not seeing.

For Owen's work came to center on denying evolution and claiming that "man is the sole species of his genus, the sole representative of his order and subclass."[53] Were this to be true, it would mean that humans had no relationship to apes or any other animal. Owen went further to deny evolutionary adaptation: "We have not a particle of evidence that any species of bird or beast that existed during the [P]liocene period has had its characteristics modified in any respect by the influence of time or of change of external influences."[54] In defense of these ideas, Owen was led into a series of debates and defeats with Thomas Huxley, in which he attempted to claim that first the teeth and then the brain of humans marked them out as distinct from all apes and simians. Owen had come to this conclusion as early as 1845, when he attempted and failed to deduce "the physiological possibility of the development of the Hottentot from the chimpanzee."[55] Like Cuvier, Owen assumed that the Khoisan peoples were the most "primitive" of humans and therefore ought to be most easily connected to the apes. If that failed, Owen "reasoned," then all such linkage failed. Racialized separation and distinction were constitutive of the highly precise visual and verbal taxonomy produced both to create remarkable new forms of knowledge from extinct birds to the ice ages and then-present-day natural history and to support white supremacy, racial hierarchy, and colonization. The place where these two taxonomies met was the line: the color line, especially the line between simian and human, and its double, the geological boundary between one era and the next.

BREAK

> The civilised races of man will almost certainly exterminate, and replace, the savage races throughout the world. At the same time, the anthropomorphous apes . . . will no doubt be exterminated. The *break* between man and his nearest allies will then be wider, for it will intervene between man in a more civilized state, as we may hope, even than the Caucasian, and some ape as low as a baboon, instead of, as now, between the negro or Australian and the gorilla.[56]

DRAWING THE LINE TODAY

If this concept of a "break" was, as it were, built in to geology, life science, and the understanding of life on Earth, and then sanctioned by Darwin himself, it continues to be active even among present-day disputes in Earth system science. Humanists have widely debated what to call the new geological era.[57] However, at the time of writing in 2015–16, an extraordinary and quite virulent debate has broken out among geologists and other Earth system scientists as to whether the Anthropocene should be bounded in time or by physical markers. As Simon Lewis and Mark Maslin, two of the leading protagonists, have observed, these issues involve "geological, philosophical and political paradigm shifts."[58] Their proposal in *Nature* for March 2015 on how to place the Global Boundary Stratotype Sections and Points, or GSSPs, colloquially known as the "golden spike" marking the stratigraphy of the Anthropocene, points directly at colonialism and imperialism.[59] When Nobel Prize–winning chemist Paul Crutzen first came up with the concept of the Anthropocene in 2000, he suggested that it began when James Watt patented the steam engine in 1784. This date has been very popular with Marxists, who are only too happy to blame the Anthropocene on what Andreas Malm has called "fossil capital."[60] While there is no denying the connection between fossil fuel, capitalism, and global warming, geologists cannot accept, Lewis and Maslin argue, the beginning of the Industrial Revolution as a golden spike both because there is no clear stratigraphic marker at that time (CO_2 levels did not notably rise in 1784 or thereabouts) and because the shift was not systemic but local to Britain. By contrast, they propose a new date of 1610, because "the arrival of Europeans in the Americas also led to a large decline in human numbers. From 61 million people, numbers rapidly declined to about 6 million people by 1650 via exposure to diseases carried by Europeans, plus war, enslavement and famine. The accompanying near-cessation of farming and reduction in fire use resulted in the regeneration of over 50 million hectares of forest."[61] While the European genocide (my term, not theirs) of the indigenous population is well known, it now appears to be coupled to a visible "golden spike" in terms of a sharp drop in atmospheric CO_2 caused by reforestation.[62] The Anthropocene began with a massive colonial genocide, in short. The implications are wide ranging. It seems no coincidence that it was in 1619 that enslaved Africans first landed in Virginia, a needed labor force in the absence of the indigenous, who themselves had served as slaves. The "virgin" forest so often lamented in its disappearance by Audubon and others had overtaken the anthropogenic grasslands and forests created by the Native populations. Depictions of the European settlement have often represented America as a virgin Native woman (as in Jan van der Straets's *Discovery of America* [Metropolitan Museum, 1587–89]), who, by implication, the "white" person was to possess in all senses. This possession has often been represented as rape. It seems that it was more like necrophilia.

For Lewis and Maslin, it should be stressed, the key to their preference for this date was not human mortality but the visible, human-caused exchange of species from viruses to plants and animals across the world that followed encounter in the Americas: "The long-term change to the Earth's trajectory is the irreversible cross-continental movement of species, and between disconnected oceans, for example, seen as maize fossil pollen at ~1600 in a European marine sediment."[63] Known to historians as the Columbian exchange, Lewis and Maslin reinforce their analysis with reference to Immanuel Wallerstein and other historians of world systems: the key definition of the Anthropocene has become cross-disciplinary in fascinating ways. Their second possibility for the golden spike is the 1964 peak in radiocarbon caused by atmospheric nuclear weapons tests. Here the imperial struggle between the United States and Union of Soviet Socialist Republics for what they perceived to be world domination was the issue. Again, however, the spike was an unintended consequence of that contest, one that was marginal to it but significant to the Earth system. Lewis and Maslin's proposal does not appear to have met with wide approval. Other suggestions are circulating. None of these scientists deny the impact of climate change. Their concern is how to define and draw the line between the human era and its predecessors. Some want to rename the Holocene as the Anthropocene, meaning that it began about twelve thousand years ago.[64] Others assert that there is no geological basis for defining a new era at all.[65]

However, the Anthropocene Working Group (AWG) of the International Commission on Stratigraphy is clearly of the opinion that the new geological era began at 05:29:21 Mountain War Time (±2s) on July 16, 1945, the instant that the U.S. military exploded a nuclear device at the Trinity site in New Mexico.[66] Whereas geological time was once seen to divide across a visible line in the material substrate of the Earth, the official proposal now intends to begin the human era with a specific moment of human activity. This type of marker, known as Global Standard Stratigraphic Age, or GSSA, is commonly used for more recent geology (measured in millions of years, it should be said). The AWG sees this as a "clear, objective moment in time" from which to date the Anthropocene. It had been widely predicted that this decision would have been taken by now, so there must be some controversy within the highest reaches of geology. Perhaps they share what seems to an outsider like a case of moving the goalposts from physical markers to moments in time. Fully anticipating such responses, supporters of the AWG argue that a "paradigm shift" has occurred from distinct sciences to the "Earth system" that understands processes as interlocked and interrelated cycles. In the opinion of Australian professor of public ethics Clive Hamilton, Lewis and Maslin have a "golden spike fetish" that prevents them from understanding the force of this shift. Hamilton argues that selecting the Trinity test "signaled unambiguously the dawn of the era of global economic domination by the United States of America, which was intimately tied to the economic boom of the post-war years and so the rapid increased [*sic*] in greenhouse gas emissions and associated warming."[67] Hamilton is referring here to what is known as the "Great Acceleration," the takeoff in CO_2 emissions and other markers of human alteration of the Earth system post-1945. As is evident both from data visualizations of the Great Acceleration and a minimal review of post-1945 history, that outcome was far from self-evident in July 1945. It shows that, however the Anthropocene is to be defined, whether by time or geological marker, interpretations of world history will be bound up with it. Scientists may need to be instructed in resisting the temptations of teleology.

Whatever the Anthropocene may be, it is not now being defined by the observation of data but by interpretation, the traditional task of the humanist. Perhaps what have been called the "posthumanities" ought also to involve the "postsciences." While geologists and Earth system scientists are making a decision as to which index to use, whether temporal or stratigraphic, the rest of us do not have to do so. Rather, we should use all the "spikes," physical and temporal, as a way to triangulate the modern, to reframe industrial capitalism, and to periodize our investigations into the Anthropocene. Unlike the AWG, we do not need to be obsessed with the split second of the GSSA, but we should consider the relay between time and geological marker as itself a new cultural formation. We should do so in the clear understanding that it is not all people that are indicted by the onset of the Anthropocene but a specific set: colonial settlers, enslavers, and would-be imperialists.

To set the Anthropocene boundary in 1610 indicates that human action was part of the transformation of the planet, but not all of it. Trees and other vegetation contributed to the drop in CO_2. The key development for Lewis and Maslin is the so-called Columbian exchange that brought different animal, plant, and virus populations into contact for the first time, with world historical and Earth system consequences. J. R. McNeill has shown that European conquest of the Americas was accidentally facilitated by the spread of viruses to which the local populations had no immunity.[68] He follows up by pointing out that late-eighteenth-century independence movements and revolutions in the Americas were also successful in part because armies sent from Europe to pacify them succumbed instead to yellow fever, malaria, and other diseases to which the hybrid American populations had acquired resistance.[69] Lewis and Maslin point to "the irreversible cross-ocean movement of species" in the Columbian exchange as a "near-permanent change to Earth."[70] Humans here act as vectors—what they do has immense consequences, but ones that they do not foresee. At the same time, this interpretation minimizes the role of colonized and enslaved peoples in that change. Enslaved Africans, for example, both brought plants with them to the Americas and knew how to cultivate them. Perhaps the most notable example was rice, long cultivated by West Africans. Forcibly removed to North America, they brought this understanding with them, as Judith Charney explains: "The development of rice culture [in America] marked not simply the movement of a crop across the Atlantic but also the transfer of an entire cultural system, from production to consumption."[71] Furthermore, the long-standing idea that viruses caused the genocide of the Native peoples of the Americas has recently been challenged by historian Andrés Reséndez, whose educated guess is that "between 1492 and 1550, a nexus of slavery, overwork and famine killed more Indians in the Caribbean than smallpox, influenza and malaria."[72] The 1610 golden spike was driven in by slavery more than it was by epidemics.

By contrast, to date the Anthropocene from the first atomic weapon is to give a certain set of humans far more deliberate power. As the director of the Manhattan Project, Robert J. Oppenheimer, put it at the Trinity site, "I am become Death, the destroyer of worlds." This Nietzschean becoming paradoxically places the future of the planet solely in human hands. As Ellen Crist has described, such supremacism is increasingly common: "In the Anthropocene discourse, we witness history's projected drive to keep moving forward as history's conquest not only of geographical space but now of geological time as well."[73] This brave new era is well captured in the title of Clive Hamilton's 2013 book *Earthmasters.* Who, though, are these "masters"? Not people as a whole, so much as an elite minority. Put bluntly, it seems that white supremacy, not content with being the übermensch, has settled on the ultimate destiny of being a geological agent.

For the AWG, choice is not just for humans over the nonhuman, as salient as it is to make this point. Lewis and Maslin have themselves pointed out the "Anthropocentrism" of the AWG's new timing.[74] It is for a certain highly privileged group of humans over all other humans and the nonhuman, as reflected in the AWG itself. Lewis and Maslin tallied the public positions of AWG members against the new collective position to show that it is in fact a minority view.[75] This objection led to a change in attribution from the electronic posting of *The Anthropocene Review* debate to its printing. The response to Lewis and Maslin is now attributed to "Members of the Anthropocene Working Group."[76] To reinforce this point, it is noticeable that all thirty-five members of the group are from North America and Europe, with the exception of one Brazilian, one South African, one Chinese, and one Kenyan scientist.[77] Only three are women, including the redoubtable Naomi Oreskes, author of the classic *Merchants of Doubt* about climate deniers and corporate funding, who is notably not a signatory to the AWG response to Lewis and Maslin. Can this group be said to be representative? Of course, Lewis and Maslin are white male scientists from the United Kingdom as well. The point is ultimately not the identity of the decision makers but the interpretation of history their choice enshrines.

TO CHANGE IT

The formation of a discourse of extinction was an entanglement of hypervision and description designed to negotiate and negate the era of abolition and revolution by generating lines of force that separated and distinguished permanent races in the human and nonhuman worlds, which tended toward extinction by virtue of their positioning in the hierarchy of what Donna Haraway has called *natureculture.* The "break" between eras was matched and reinforced by the "breaks" between races. To envisage the Anthropocene as the white supremacy scene is, then, simply to articulate its own logic, recently defined by Hamilton as "significant human disturbance to the functioning of the Earth system."[78] What is required is not just an analysis but a politics that challenges such hierarchy, as in the occupation of the Museum of Natural History on the fiftieth anniversary of the assassination of Malcolm X by #BlackLivesMatter activists. Following a tour of the museum for young Muslim women led by Malcolm himself in 1961, #BlackLivesMatter activists showed in their anniversary tour how the museum structures certain forms of interpretation of "nature" that favor white supremacy. Art historian Yates McKee narrates:

> The tour concluded with a poetic ritual at the Hall of North American Forests. The centerpiece display of the hall is a majestic 1400 hundred–year old tree ring from a giant Sequoia chopped down by loggers in 1891 and annotated by the museum with series of markers pertaining to the supposedly universal human history that had unfolded since the tree first began to grow. Against the grain of what she described as the "naturalization of history," the tour guide [Cherrell Brown] used the tree-ring as an opportunity to meditate on the etymological resonance between the words "root" and "radical."[79]

After reprising the Yoruba greeting and praise word *ashe,* or "life-force," activists fixed a banner reading "White Supremacy Kills" to the statue of Theodore Roosevelt that dominates the entrance to the museum. Indeed, the Museum of Natural History was an outspoken supporter of eugenics in the early twentieth century, visually depicted in the statue as Roosevelt towers over half-naked figures of an African and a Native.

In her account of climate change, activism, and capitalism, *This Changes Everything,* Naomi Klein concludes by drawing parallels between the abolition of slavery and the struggle against fossil fuel capitalism.[80] It seems exactly the right comparison to make. It implies considering what kind of activism is required to end fossil fuel dominance. Here I have argued that recognizing the Anthropocene as part of the structures of white supremacy is a key first step. The action in the Natural History Museum shows that antiracism challenges presumed allies within the climate movement and the usual capitalist suspects alike. Until and unless this deficiency is addressed, with all its ancillary questions from pollution exposure to prison ecology, there will not be a climate movement of sufficient force to challenge the current interaction of fossil fuels, capitalism, and limited democracy. At the same time, Klein's parallel raises a second question. In his classic *Black Reconstruction,* W. E. B. Du Bois showed that slavery fell because of the "general strike" of the enslaved, not because of Northern abolitionism. Half a million of the enslaved upped and left the Confederacy, ending slavery by default two years before the Emancipation Proclamation in 1863. Those of the strikers who joined the Union armies are now accepted to have helped swing the Civil War against the South. It is fashionable among white Marxists of a certain age to say that it is harder to imagine the end of capitalism than the end of the world. On the contrary, it is far easier to envisage a mass resistance of those designated not human by white supremacy against fossil fuel capitalism. Slavery was ended by the transnational resistance of the enslaved that grew from local actions to the Haitian revolution of 1791 and the general strike against U.S. slavery. The year before Haiti's uprising, ending slavery was as improbable as ending fossil fuel capitalism sometimes seems today. It is up to all of us to see that history repeats itself, not as tragedy or farce, but as the sequel that is better than the original.

NOTES

1. Simon Lewis and Mark Maslin, "Defining the Anthropocene," *Nature* 519, no. 7542 (2015): 172.

2. To quote the 2001 Amsterdam Declaration of Earth System Science, "The Earth system behaves as a single, self-regulating system comprised of physical, chemical, biological and human components." International Geosphere-Biosphere Programme, *2001 Amsterdam Declaration of Earth System Science,* July 13, 2001, http://www.igbp.net/about/history/2001amsterdamdeclarationonearthsystem science.4.1b8ae20512db692f2a680001312.html.

3. Joe Feagin and Sean Elias, "Rethinking Racial Formation Theory: A Systemic Racism Critique," *Ethnic and Racial Studies* 36, no. 6 (2016): 931–60.

4. See Prison Ecology Project, http://nationinside.org/campaign/prison -ecology/.

5. See Kavita Philip, "Doing Interdisciplinary Asian Studies in the Age of the Anthropocene," *Journal of Asian Studies* 73, no. 4 (2014): 975–87; Elizabeth Johnson and Harlan Morehouse, "After the Anthropocene: Politics and

Geographic Inquiry for a New Epoch," *Progress in Human Geography* 38, no. 3 (2014): 439–56; Phoebe Godfrey, "Introduction: Race, Gender and Class and Climate Change," *Race, Gender, and Class* 19, no. 1/2 (2012): 3–11.

6. Clive Hamilton, *Earthmasters: The Dawn of the Age of Climate Engineering* (New Haven, Conn.: Yale University Press, 2013).

7. See Marisol de la Cadena, "Uncommoning Nature," *eflux journal* 65 (May–August 2015), http://supercommunity.e-flux.com/texts/uncommoning-nature/.

8. Donald Moore, Jake Kosek, and Anand Pandian, "Introduction: The Cultural Politics of Race and Nature: Terrains of Power and Practice," in *Race, Nature, and the Politics of Difference* (Durham, N.C.: Duke University Press, 2003), 3.

9. Scott v. Sandford, 60 U.S. 393, 406–7, https://www.law.cornell.edu/supremecourt/text/60/393#writing-USSC_CR_0060_0393_ZO.

10. Bruno Latour, quoted in Jane Bennett, *Vibrant Matter: A Political Ecology of Things* (Durham, N.C.: Duke University Press, 2010), 109.

11. Bennett, *Vibrant Matter,* 111.

12. Hardin, quoted ibid., 27.

13. Dipesh Chakrabarty, "The Climate of History: Four Theses," *Critical Inquiry* 35 (Winter 2009): 211.

14. Centers for Disease Control and Prevention, "Deaths Associated with Hurricane Sandy—October–November 2012," *Morbidity and Mortality Weekly Report* 62, no. 20 (2013): 393–97, https://www.cdc.gov/mmwr/preview/mmwrhtml/mm6220a1.htm.

15. Jacques Rancière, *Hatred of Democracy* (New York: Verso, 2006), 62.

16. C. L. R. James, *The Black Jacobins* (1938; repr., London: Allison and Busby, 1968).

17. Eric Williams, *Capitalism and Slavery* (1944; repr., Raleigh: University of North Carolina Press, 1994).

18. Edward E. Baptist, *The Half Has Never Been Told: Slavery and the Making of American Capitalism* (New York: Basic Books, 2014), xix. See also Walter Johnson, *River of Dark Dreams: Slavery and Empire in the Cotton Kingdom* (Cambridge, Mass.: Belknap Press, 2013).

19. Catherine Hall, Nicholas Draper, Keith McClelland, Katie Dinington, and Rachel Lang, introduction to *Legacies of British Slave-Ownership: Colonial Slavery and the Formation of Victorian Britain* (Cambridge: Cambridge University Press, 2014), 4.

20. Richard H. Grove, *Green Imperialism: Colonial Expansion, Tropical Island Edens and the Origins of Environmentalism* (Cambridge: Cambridge University Press, 1995), 71ff.

21. Denise Ferreira da Silva, *Toward a Global Idea of Race* (Minneapolis: University of Minnesota Press, 2007), xiii.

22. Karl Marx, chapter 7 in *The Eighteenth Brumaire of Louis Bonaparte,* originally published in the first issue of *Die Revolution,* 1852, https://www.marxists.org/archive/marx/works/1852/18th-brumaire/ch07.htm.

23. Da Silva, *Toward a Global Idea of Race,* xxxix.

24. Charles Lyell, *Principles of Geology; or, the Modern Changes of the Earth and Its Inhabitants, Considered as Illustrative of Geology* (1830; repr., London: John Murray, 1840), 1.

25. For a comprehensive history of geology in the period, see Martin Rudwick, *Bursting the Limits of Time: The Reconstruction of Geohistory in the Age of Revolution* (Chicago: University of Chicago Press, 2005).

26. "Every organized being reproduces others that are similar to itself, otherwise, death being a necessary consequence of life, the species would become extinct." Georges Cuvier, *The Animal Kingdom: Arranged in Conformity with Its Organization,* trans. H. M'Murtrie (1812; repr., New York: Carvill, 1832), 1:17.

27. Ibid., 1:50. This translation is more violently racist than the one used most often today. I cite it because it was the one in circulation in the United States in the period—the copy digitized by Google comes from Harvard Libraries.

28. Georges Cuvier, *Recherches sur les ossemens fossiles,* vol. 1 (Paris: Deterville, 1812), quoted in Stephen Jay Gould, *The Mismeasure of Man,* rev. ed. (London: Penguin, 1996), 69.

29. Richard Rhodes, *John James Audubon: The Making of an American* (New York: Alfred A. Knopf, 2004), 4–5.

30. John James Audubon, *Writings and Drawings,* ed. Christopher Irmscher (New York: Library of America, 1999), 47.

31. Linda Dugan Partridge, "By the Book: Audubon and the Tradition of Ornithological Illustration," *Huntington Library Quarterly* 59, no. 2/3 (1996): 269–301.

32. Audubon, *Writings and Drawings,* 261.

33. Christopher Iannini, *Fatal Revolutions: Natural History, West Indian Slavery, and the Routes of American Literature* (Raleigh: University of North Carolina Press, 2012), 261.

34. John James Audubon, *Ornithological Biography; or, An Account of the Habits of the Birds of the United States of America* (Edinburgh: Adam Black, 1831–39), 2:173, 178.

35. John James Audubon, "The Runaway," in *Ornithological Biography,* 2:27–32.

36. Sylviane A. Diouf, *Slavery's Exiles: The Story of the American Maroons* (New York: New York University Press, 2014), 87.

37. Agassiz, quoted in Christoph Irmscher, "Agassiz on Evolution," *Journal of the History of Biology* 37, no. 1 (2004): 205–7.

38. Edouard Desor, *Excursions et séjours dans les glaciers et les hautes régions des Alpes de M. Agassiz et ses compagnons de voyage* (Neuchâtel: J.-J. Kissling, 1844), 10.

39. Louis Agassiz, "Discours," in *Actes de la Société helvétique des sciences naturelles, réunie à Neuchâtel les 24, 25, et 26 juillet 1837, 22e session* (Neuchâtel: Imprimerie de Petitpierre, 1837), xxxi.

40. Louis Agassiz, *Etudes sur les glaciers* (Neuchâtel: Ol. Petitpierre, 1840), 19.

41. For history being deduced like faint traces on a lithographic stone, see Agassiz, *Etudes,* 241; for the fossilized photograph, see Joanna Zylinska's chapter in this volume (chapter 3).

42. Agassiz, *Etudes,* 241.

43. Thomas Carlyle, *The French Revolution* (1837; repr., Boston: Dana Estes, 1892), 2:46.

44. Ibid., 2:219.

45. Niles Eldredge and S. J. Gould, "Punctuated Equilibria: An Alternative to Phyletic Gradualism," in *Models in Paleobiology,* ed. T. J. M. Schopf, 82–115 (San Francisco: Freeman Cooper, 1972).

46. Stephen Jay Gould, *The Mismeasure of Man* (New York: W. W. Norton, 1981), 74–82.

47. Agassiz, quoted in Molly Rogers, *Delia's Tears: Race, Science, and Photography in Nineteenth-Century America* (New Haven, Conn.: Yale University Press, 2010), 281.

48. Ingraham, quoted in Walter Johnson, *Soul by Soul: Life inside the Antebellum Slave Market* (Cambridge, Mass.: Harvard University Press, 1999), 135. Johnson's chapter 5, "Reading Bodies and Marking Race," is indispensable for understanding how "blackness" was legible under slavery (135–61).

49. Rogers, *Delia's Tears,* 219.

50. Alan Trachtenberg, *Reading American Photographs: Images as History—Mathew Brady to Walker Evans* (New York: Hill and Wang, 1990), 60, cited in Molly Rogers, "Fair Women Are Transformed into Negresses," Mirror of Race, January 18, 2012, http://mirrorofrace.org/fair-women/#4a.

51. Brian Wallis, "Black Bodies, White Science: Louis Agassiz's Slave Daguerreotypes," *American Art* 9, no. 2 (1995): 38–61.

52. Richard Owen, *Memoirs on the Extinct Wingless Birds of New Zealand with an Appendix on Those of England, Australia, Newfoundland, Mauritius and Rodriguez,* 2 vols. (London: John Van Voorst, 1879), 1:73.

53. Richard Owen, *On the Classification and Geographical Distribution of the Mammalia* (London: John W. Parker, 1859), 103.

54. Owen, *Memoirs,* 202.

55. Richard Owen, quoting a review by Richard Owen Sr., *The Life of Richard Owen by His Grandson the Rev. Richard Owen* (London: John Murray, 1894), 251.

56. Emphasis added. Charles Darwin, *The Descent of Man* (London: John Murray, 1871), 1:172–73, quoted in da Silva, *Toward,* 110.

57. Donna Haraway, "Anthropocene, Capitalocene, Plantationocene, Chthulucene: Making Kin," *Environmental Humanities* 6 (2015): 159–65.

58. Mark A. Maslin and Simon L. Lewis, "Anthropocene: Earth System, Geological, Philosophical and Political Paradigm Shifts," *The Anthropocene Review* 2, no. 2 (2015): 108–16.

59. Simon L. Lewis and Mark A. Maslin, "Defining the Anthropocene," *Nature* 519 (March 12, 2015): 171–80.

60. Andreas Malm, *Fossil Capital: The Rise of Steam Power and the Roots of Global Warming* (New York: Verso, 2016).

61. Lewis and Maslin, "Defining," 175.

62. See Charles C. Mann, *1491: New Revelations of the Americas before Columbus* (New York: Vintage, 2006).

63. Simon L. Lewis and Mark A. Maslin, "A Transparent Framework for Defining the Anthropocene Epoch," *The Anthropocene Review* 2, no. 2 (2015): 134.

64. Bruce D. Smith and Melinda A. Zeder, "The Onset of the Anthropocene," *Anthropocene* 4 (December 2013): 8–13.

65. Mike Walker, Phil Gibbard, and John Lowe, "Comment on 'When Did the Anthropocene Begin? A Mid-Twentieth Century Boundary Is Stratigraphically Optimal' by Jan Zalasiewicz et al. (2015), *Quaternary International,* 383, 196–203," *Quaternary International* 383: 204–7.

66. Jan Zalasiewicz, Colin N. Waters, Mark Williams, Anthony D. Barnosky, Alejandro Cearreta, Paul Crutzen, Erle Ellis et al., "When Did the Anthropocene Begin? A Mid-Twentieth Century Boundary Is Stratigraphically Optimal," *Quaternary International* 383 (October 5, 2015): 196–203.

67. Clive Hamilton, "Getting the Anthropocene So Wrong," *The Anthropocene Review* 2, no. 2 (2015): 104.

68. J. R. McNeill, *Mosquito Empires: Ecology and War in the Greater Caribbean 1620–1914* (Cambridge: Cambridge University Press, 2010).

69. McNeill is a signatory to the Anthropocene Working Group's rebuttal to Lewis and Maslin, so he might not approve of the use of his work in this context. See note 76.

70. Lewis and Maslin, "A Transparent Framework," 144.

71. Judith A. Charney, *Black Rice: The African Origins of Rice Cultivation in the Americas* (Cambridge, Mass.: Harvard University Press, 2002), 2.

72. Andrés Reséndez, *The Other Slavery: The Uncovered Story of Indian Enslavement in America* (New York: Houghton Mifflin Harcourt, 2016), 17.

73. Eileen Crist, "The Poverty of Our Nomenclature," *Environmental Humanities* 3 (2013): 132.

74. Lewis and Maslin, "Anthropocene," 109.

75. Lewis and Maslin, "A Transparent Framework," 143.

76. Members of the Anthropocene Working Group: Jan Zalasiewicz, Colin N. Waters, Anthony D. Barnosky, Alejandro Cearreta, Matt Edgeworth, Erle C. Ellis, Agnieszka Gałuszka, Philip L. Gibbard, Jacques Grinevald, Irka Hajdas, Juliana Ivar do Sul, Catherine Jeandel, Reinhold Leinfelder, J. R. McNeill, Clément Poirier, Andrew Revkin, Daniel deB Richter, Will Steffen, Colin Summerhayes, James P. M. Syvitski, Davor Vidas, Michael Wagreich, Mark Williams, and Alexander P. Wolfe, "Colonization of the Americas, 'Little Ice Age' Climate,

and Bomb-Produced Carbon: Their Role in Defining the Anthropocene," *The Anthropocene Review* 2, no. 2 (2015): 117–27.

77. Subcommission on Quaternary Stratigraphy, "Working Group on 'the Anthropocene,'" Subcommission on Quaternary Stratigraphy, http://quaternary.stratigraphy.org/workinggroups/anthropocene/.

78. Clive Hamilton, "The Anthropocene as Rupture," *The Anthropocene Review* 3, no. 2 (2016): 102.

79. Yates McKee, *Strike Art: Contemporary Art and the Post-Occupy Condition* (New York: Verso, 2016).

80. Naomi Klein, *This Changes Everything: Capitalism vs. the Climate* (New York: Simon and Schuster, 2014), 455–57.

7

Lives Worth Living: Extinction, Persons, Disability

Claire Colebrook

What is the relationship between extinction and disability? One of the ways in which we might think about disability and disability studies is as requiring an expansion of conditions of justice; this is how Martha Nussbaum has criticized the liberal tradition of fairness and personhood in her book *Frontiers of Justice: Disability, Nationality, Species Membership*. We should, she argues, extend considerations of fairness to include those who care for others. If we think about a world that enables human capacities and flourishing, then we need to look beyond autonomous and self-defining individuals. Disability considerations would both enhance and extend the range of political compassion, enabling a notion of persons that is not merely that of the abstract political subject but a being with capacities and dignity; capacities are richer and more varied than our narrow notion of personhood currently allows. For Nussbaum, we will live in a better world if we expand our notion of capacity and what counts as a flourishing human life.[1]

In what follows, I want to reverse this relation, and rather than expand capacities and justice to allow for disability (with disability being the secondary consideration), I want to see disability as the primary or transcendental condition from which the supposedly "normal" person derives, *and* further claim that the long history of the "normal" subject is directly intertwined with the accelerated extinction of humans and nonhumans. If one considers the subject of capacities from which Nussbaum begins her critique—the liberal person, blessed with reason, autonomy, "favorable" social conditions, and an enlightened milieu of political deliberation—one would need to recognize the long history of enslavement (of humans and nonhumans), exploitation, appropriation, and colonization that made even the thought of the just society possible. Disability is not an added-on concern but is precisely what orients, if silently, the problem of extinction. One might say that "human" existence is constitutively disabled (or, to follow Bernard Stiegler, that its default condition is dependence on a broad network of technologies and archives that have never been equally distributed).[2] Furthermore, the capacities that enable the "able" person have cost, and continue to cost, Earth. Those lives that are (to borrow from Nick Bostrom) "technologically immature" may perhaps not be lamentable and to be avoided at all costs but perhaps offer a trajectory for life that is not necessarily that of extinction.[3]

Even though the specific concepts of extinction and disability are rarely linked explicitly, the two concepts are inextricably intertwined in discussions of what counts as a life worth living. Indeed, the grand Socratic notion that the unexamined life is not worth living is not only normative (which is almost unavoidable) but *normalizing*: to privilege the life of examination is to open up a history that will generate the individual, reflective, deliberative, and rational subject, but to make a claim about a life *not worth living* is to hint at the long history that will extinguish, eliminate, harness, and evaluate unworthy lives, and will do so precisely by way of capacity. Outside explicit work on extinction and outside the rich field of disability studies, it is possible to find constant and complex linkages between the question of the worth of life (its capacity or ability) and whether such a life ought to exist. Many such arguments are utilitarian; and while utilitarianism might seem to be but one branch of (analytic) philosophy, part of my argument will be that as a conception of the liberal subject of capacity gains ascendency and takes on increasing value in neoliberal arguments for autonomy, *and* as the planet faces accelerated and mass extinction, a utilitarian logic becomes increasingly dominant.

Utilitarianism is a motif that will necessarily haunt questions of extinction and capacity: as resources and the capacity to survive become threatened, decisions will need to be made regarding the worth of life.

Precisely in this respect, it is utilitarianism that has also articulated the most *offensive* position on disability. By "offensive," here, I am not referring to an affect or emotion but rather—as in the manner of a military offensive—to a direct and forthright targeting of what has been set aside as "disabled." Here it might seem that a utilitarian approach is partial and that there are other ethical paradigms, which of course there are, but I want to argue that the extreme positions that utilitarianism has yielded bring to the fore what is implicit in a broader history of ethics focused on personhood and a life worth living. One of the objections to calculations of utility would be by way of a deeper or inviolable conception of the person, but this too relies on distinguishing between what counts as "utility" and what would warrant a mode of "dignity" beyond calculation. For Nussbaum, the key stakes of justice lie in considering what counts as a dignified life, where dignity includes capacities that extend beyond social utility and mutual advantage. Her claim is that *dignity* should be the basis for social entitlements and that we attribute dignity not for rational and active powers but for "our" animal fragility: "bodily need, including the need for care, is a feature of our rationality and our sociability; it is one aspect of our dignity, then, rather than something to be contrasted with it."[4] This is perhaps why Nussbaum's title refers to "species membership," as though feeling and caring for one's kind (which would, in part, include nonhuman animals) are not only a *recognition* of dignity but dignify one's own life. To suffer, to be fragile, is to possess a life worth living. Here Nussbaum refers to the value and enhancement (beyond *strict* utility) of caring for others and of having social relationships with those whose capacities are not those of the classic rational individual. Her approach on capacities "includes the advantage of respecting the dignity of people with mental disabilities and developing their human potential, whether or not this potential is socially 'useful' in the narrower sense. It includes, as well, the advantage of understanding humanity and its diversity that comes from associating with mentally disabled people on terms of mutual respect and reciprocity."[5]

Nussbaum presents her account as a broadening of theories of human justice by way of a more classical conception of the life worth living, one not reduced to narrow notions of mutual advantage. Even though her

discourse and disciplinary terrain might appear to be strictly philosophical, the very mode of posing the question of *what we owe to a life* is really
(ultimately) the question that presses itself upon human civilization now,
and always. As "we" look to the future and the sixth great extinction event,
the question of who and what survives will be imposed upon us. Utilitarian approaches to this question are, as I have already suggested, *offensive,*
but they are because they disclose something offensive—or combative,
violent, conquering—in the philosophical tradition of dignified humanity and the life worth living. In this respect, disability is neither a recent
nor a local concern: the very formation of the Greek polity is based on
the exclusion of those with lesser capacities. Even though, as Lennard
Davis has argued, the notion of the "normal" body is very recent and
is quite different from earlier cultures' conception of an *ideal* body that
no actual member of the species achieves, the exclusion of those who
do not possess the proper potentiality of political humanity has been at
the basis of the history of the Western polity.[6] When Nussbaum argues
for an expanded sense of capacities, she nevertheless, and necessarily,
maintains the question of the life *worth living.* This classic philosophical question always and necessarily invokes ability, or, more accurately,
disability, and this in two respects. Not only are subjects defined by way
of powers (of reason, deliberation, and empathy) but those capacities
in turn are enabled by a history of technologies and archives on which
"able" subjects are increasingly dependent. At the very least, definitions
of proper political persons rely on quite specific capacities that, even in
expanded scenarios, are not all-inclusive. More importantly, the quite
specific concept of the liberal, deliberative, rational, and empathetic
subject depends on a history of "enlightenment" that disabled many lives,
by way of exclusion, colonialism, resource depletion, or expropriation.
In a world where not all lives matter to the same extent, the concept
of disability is precisely what enables political inclusion, privilege, and
personhood. When Peter Singer argues, in a manner that appears to be
exceptional, and exceptionally offensive, that rationality and autonomy
(and not species membership) are the capacities that would preclude us
from being right in killing another human being, he is taking part in a far
broader offensive that is definitive of the philosophical epoch oriented

around the question of the life worth living.[7] For Singer, what matters when considering a life and its worth is not that life's capacities but its capacity to suffer; however, this nevertheless raises the worthiness and power of affect. What has proven to be so shocking about Singer's work is his highlighting of a rationality already at work in claims regarding human dignity: we have already deployed notions of value and worthiness, values that Singer wishes to shift from species membership or supposed rational powers to affective powers. Not only is the question of the life worth living offensive (in its implicit generation of an unworthy life) but the life *worth living* is—for all its rhetoric of autonomy and power—a life of *dependence and incapacity,* generated through a history of enlightenment that is a history of appropriation, plundering, brigandry, excessive consumption, and energy profligacy. The Cartesian reflective subject is utterly dependent on networks of labor and technology that bolster his power while remaining outside immediate command; and as the history of enlightenment progresses, so does felicitous incapacity. "We" become more and more powerful by way of networks—the Internet, data, cheap goods, cheap skills—that rely on others' capacities. Our exceptional political ability as subjects of reason is twinned with intensified incapacity, just as our autonomy is ultimately dependent on a history of ongoing slavery. Could we have the able political subject of deliberation and reason without the planet-destructive history of industrialism and globalism that at once enables and disables what has come to be known as humanity? Could there have been a tradition of "the life worth living" without a global industry that generated unworthy and dis-abled lives? And is not the question of the life worth living, the capable life, intertwined essentially with dependence and incapacity?

What I want to question here is whether such a question can have any coherence at all in an epoch of extinction: to ask about lives worth living is necessarily to be *offensive,* valuing the worth of some lives over others, and thereby waging violence (however slow) against some forms of life. If, as I would also argue, any epoch of thriving and fecundity takes place at the expense of some lives, then all ages are ages of extinction. What makes our time—the sixth mass extinction—more *intense* is that questions that have always haunted political personhood are now becoming more explicit. The interrelated problem of capacity and extinction has not only determined the human lives that are deemed to be worth living but has also generated the liberal political person whose autonomy, productivity, super-intelligence, and heightened capacity for urbanity is the Anthropos of the Anthropocene, the "man" whose cost to the planet is too exorbitant to reckon.[8] When (today) utilitarian arguments are explicitly offensive, or make the claim that some lives ought not be lived, they reveal the *offensive* (combative, polemical, violent, barbaric, sacrificial) nature of what has called itself civilization. If this civilization, today, is facing extinction and is therefore pressed—more than ever—to consider ways of "weighing lives," it may either continue with ever more nuanced and expanded conceptions of the worth of life, or it may regard this question itself as an indictment of the very rationality it seeks to save. Phrased differently, we might say that the problem of disability runs to the very heart of the extinction-logic that enables the political tradition of the person. Both those who assume that the human species—because of certain capacities—has a prima facie right to survive *and* those who calculate that human life as such is not worth living (for all their seeming extremity) are expressing a broader logic of the proper potentiality of a highly normative conception of human flourishing. As an example of the prima facie "right to humanity," I would cite Rebecca Newberger Goldstein's defense of Wilfrid Sellars and philosophical progress. The rational image we have of ourselves, even when at odds with scientific evidence about the irrational causes of our behavior, will generate an ongoing history of *coherence* and *inclusion,* where the rational "we" extends itself to value others:

> Gregarious creatures that we are, our framework of making ourselves coherent to ourselves commits us to making ourselves coherent to others. Having reasons means being prepared to share them—though not necessarily with everyone. The progress in our moral reasoning has worked to widen both the kinds of reasons we offer and the group to whom we offer them. There can't be a widening of the reasons we give in justifying our actions without a corresponding widening of the audience to which we're prepared to give our reasons. Plato gave arguments for why Greeks, under the pressures of war, couldn't treat other Greeks in abominable ways, pillaging and razing their cities and taking the vanquished as slaves. But his reasons didn't, in

principle, generalize to non-Greeks, which is tantamount to denying that non-Greeks were owed any reasons. Every increase in our moral coherence—recognizing the rights of the enslaved, the colonialized, the impoverished, the imprisoned, women, children, LGBTs, the handicapped . . .—is simultaneously an expansion of those to whom we are prepared to offer reasons accounting for our behavior. The reasons by which we make our behavior coherent to ourselves changes together with our view of who has reasons coming to them.

And this is *progress,* progress in increasing our coherence, which is philosophy's special domain. In the case of manumission, women's rights, children's rights, gay rights, criminals' rights, animal rights, the abolition of cruel and unusual punishment, the conduct of war— in fact, almost every progressive movement one can name—it was reasoned argument that first laid out the incoherence, demonstrating that the same logic underlying reasons to which we were already committed applied in a wider context. The project of rendering ourselves less inconsistent, initiated by the ancient Greeks, has left those ancient Greeks, even the best and brightest of them, far behind, just as our science has left their scientists far behind.

This kind of progress, unlike scientific progress, tends to erase its own tracks as it is integrated into our manifest image and so becomes subsumed in the framework by which we conceive of ourselves.[9]

For all its manifest worthiness, the notion of a progressive "self-image" that gains in ongoing global coherence, alongside scientific progress, sees its path of self-correction as improving with more and more human life taking part in the journey of development. One could make the rather obvious point that such a notion of "progress" by way of inclusion and ongoing "self-image" precludes other ways of thinking about human and nonhuman life that do not involve self-image (or some shared normative conception of "the human"), but in addition to the colonialist mentality of self-justification, one might ask about the price paid for such a history of philosophical progress. Would not other modes of life—such as those without an overinvestment in "self-image" or "the" human—have generated a quite different history of the planet? Such a question cannot be asked if a certain mode of human reason is an unquestioned good. But just as the inflation of human personhood precludes asking the question of the loss and extinction of other lives with other capacities, certain arguments for the extinction and annihilation of part or all of humanity also assume the value of the person—a single life with its specific coherence, value, and meaning. (Not only is such a notion historically and culturally specific, and tied to a highly normative conception of human self-awareness, it is also *this* self with an unquestioned right to the "good life" of reflection, reason, and self-determination that has generated the Anthropocene.)

When this prima facie right to life has been questioned, it has, more often than not, been by way of the same norms of capacity, will, autonomy, and personhood that supposedly make life worth living. David Benatar has argued that the human species as such should—after rational consideration—decide that it ought not exist. If we were to calculate the pleasures and pains of human existence, then not only would we decide on nonexistence as the best way to ensure the reduction of suffering, we would also realize that while there is an imperative to eliminate suffering, there is no symmetrical imperative to bring persons into being to generate pleasures or well-being. Benatar does not see a performative contradiction in being a will who decides that it is better not to exist as a willing being; once we come into being, there is a rational reason to persist in our being and live as well as possible, but that does not entail that we should will other lives to come into being. Benatar's argument is an intensified form of an argument that has profound implications for disability.[10] Peter Singer has argued that being human is not sufficient to justify a life worth living and that the calculus of pain, suffering, and living well should prompt us to choose the lives of some animals—who could enjoy lives free of suffering—over the lives of some humans, whose quality of life would not count as living well. It is for this reason that Singer can at once argue that animals ought not be killed for human consumption and that some forms of infanticide are legitimate. For Singer, it is the lack of rationality, autonomy, and a certain appreciation of life (rather than being human) that renders life not worth living:

> The fact that a being is a human being, in the sense of a member of the species *Homo sapiens,* is not relevant to the wrongness of killing it; it is, rather, characteristics like rationality, autonomy, and self-consciousness that make a difference. Infants lack these characteristics. Killing them, therefore, cannot be equated with killing normal human beings, or any other self-conscious beings.[11]

Singer expands on this point by considering a specific type of disability and what it precludes:

> To have a child with Down syndrome is to have a very different experience from having a normal child. It can still be a warm and loving experience, but we must have lowered expectations of our child's ability. We cannot expect a child with Down syndrome to play the guitar, to develop an appreciation of science fiction, to learn a foreign language, to chat with us about the latest Woody Allen movie, or to be a respectable athlete, basketballer or tennis player.[12]

This degree of disability does not necessarily warrant infanticide or abortion, but what *does* count is development; the more capacity a being develops, the less ethical it is to terminate a life. If parents choose to abort an "abnormal" fetus, then they do so at a stage prior to the development of the capacities that would make killing unethical; the same applies to infanticide. It is not species membership but capacity that counts.

Both Benatar and Singer rely on a strict utilitarianism; species and sentiment aside, one should decide on whether a life is worth living *in general,* where worthiness can (at the very least) be determined by an absence of suffering. In contrast with arguments that begin from the sanctity of the person, one begins with a calculus: a good life is a free self-determining life. If one accepts the premise of a *life worth living,* then certain lives become candidates for nonbeing. (For Singer, this is the profoundly disabled, whereas for Benatar, it is humanity as such.) It seems that questions of utility, or of what counts as a life with a sufficient degree of pleasure (or meaningfulness, or autonomy), lead inevitably to questions of human nonbeing: are there some lives that simply should *not be?* One might respond to this by objecting that the calculus of decision presupposes that which it claims to have justified; the subject who is doing the calculating, who is deciding on what ought to survive and how lives ought to be weighed, is—needless to say—a certain type of subject. This subject has the following capacities: a sense of "a" life, a sense of *capacity* (with rationality and autonomy being of significant importance), a sense of "humanity" as a global whole of which one is a member, and a manner of looking at life in terms of worthiness. One should not need too much training in anthropology, history, or critical race studies to discern the highly specific nature of these capacities.

This is not just to make a point about the poverty and brutality of Western reason and its normalizing gestures; it is also to say that many of the critiques of that same universal subject—such as those who argue for the worth of other lives, or those who value life as such *for whatever reason*—nevertheless take part in a rationing of life that is *offensive*. Here I draw again on the necessarily offensive/combative character of any assessment of the worth of life. Even if the worth of life is defined by less strictly utilitarian categories such as "meaning" or "dignity," a certain capacity for calculus, for considering something like human life as such, and then the value of "a" life, allows for the claim that certain lives might be justifiably extinguished *and* enables a life of high capacity (high production, high reason, high technology) that has precipitated the sixth mass extinction.

The calculations of Singer and Benatar are different in important ways and related in important ways. For Benatar, a lot depends on pleasure and pain not being symmetrical: even if most of my life were one of enjoyment, the nonbeing of enjoyment is not a loss, whereas the being of suffering is a loss. Not existing, and therefore the absence of pleasure, is not a straightforward negative in the way that suffering is: when one is suffering, it makes sense to want to eliminate suffering, to will suffering away. But it does not make the same sense, in a state of nonbeing, to will pleasure (and the existence it would require) into being. Singer, by contrast, is concerned with nonbeing not because he deems human life to be worthless but because—quite the contrary—he accepts a certain worthiness of some modes of existence. There are some forms of human life that are so impoverished or incapacitated that "we" who exist and have developed reason are permitted not to bring them into being: "Shakespeare's image of life as a voyage is consistent with the idea that the seriousness of taking life increases gradually, parallel with the gradual development of the child's capacities that culminate in its life as a full person."[13]

The unit of life by which we calculate who lives and who dies (what counts as suffering) informs the question, Should we really be able to decide that some lives (ranging from all human life to disabled human lives) ought not exist? One could say, following Kant, that being able to make such a calculus—being able to ask about what life ought to be—destroys any unit that would allow lives to be weighed in relation to each other. Rather than have a measure that would negotiate who lives, one would value life precisely *because* it is without measure. Indeed, our lament or preliminary mourning for the possible extinction of humans would lie in the anticipated loss not of our species being but of the intelligence that enabled the thought of our species being. Even a cursory glance at "end-of-world" narratives reveals that what presents itself as the end of "the" world is really the end of the "rational" world of capable persons. Postapocalyptic scenarios present humans wandering aimlessly in resource-deprived landscapes, subjected (once again) to tribalism, despotism, and the loss of all "reason." (As one recent example, one might think of *Mad Max: Fury Road* [2015], where the remaining populace has become nothing more than a multitude focused on mere survival. One feature of "post"-apocalyptic cultural production is that there is a world after the end of the world, but it is no longer the world of liberal affluent personhood; "we" are suddenly "all" living in third world conditions.) One might say that what would be lost in the end of the world—or that what we fear when we contemplate human extinction—is not the loss of the world, or of life (for both would continue), but the loss of what has come to count as "rational" or "intelligent" life. It is not so much calculated as *calculating* life that is worthy of living on, and while there are some general preliminary mourning rituals for the sixth *mass* extinction, cultural production seems to be more concerned with the extinction of Western middle-class urban capitalist life. One can think here of the large number of "end-of-world" narratives that are really "end-of-Manhattan" plots, from *The Day after Tomorrow* (2004) and *Cloverfield* (2008) to the book and documentary *The World without Us* (2007), which begins by describing New York going through a slow decay after humans are no longer there to maintain the altered landscape. So, yes, there is a broad perception of the looming extinction of more than human life, but it occurs in a context of an increasing focus on the loss of the only life *worth* saving, a life that is not calculable precisely because it is the life of the point of view of reason, where reason—in turn—is a highly specific (or species-defining) range of capacities.

For Nick Bostrom (director of the Future of Humanity Institute at

Oxford University), it is obvious, upon rational reflection, that the *loss of intellectual life as such* would be of a catastrophic order that far outweighs the tragedy of losing some or many humans. Bostrom follows Derek Parfit in "demonstrating" that a loss of *all* rational human life, despite first assumptions, would be far, far worse than losing nearly all rational life.[14] Despite our first intuitions, events that appear to be profoundly catastrophic (such as the Holocaust) are—ultimately—events from which "we" recover. Truly disastrous would be a loss of rationality, rather than the loss of a very large number of humans. Bostrom calculates that most of our efforts ought to be directed at the reduction of existential risk; minimizing the risk of the catastrophic loss of intelligence *in general* is a far greater priority (or ought to be) than, say, reducing the risk of local catastrophes (such as the genocidal losses that humans have already sustained but which, on reflection, do not amount to that much of a loss in the scheme of things). So we might say that both for Benatar and for Bostrom, despite the seemingly opposed claims *for* human extinction (Benatar) or human survival at all costs (Bostrom), there is a prima facie value placed on human capacity defined as rationality of a certain mode:

> If we suppose with Parfit that our planet will remain habitable for at least another billion years, and we assume that at least one billion people could live on it sustainably, then the potential exists for at least 10^{16} human lives of normal duration. These lives could also be considerably better than the average contemporary human life, which is so often marred by disease, poverty, injustice, and various biological limitations that could be partly overcome through continuing technological and moral progress.
>
> However, the relevant figure is not how many people could live on Earth but how many descendants we could have in total. One lower bound of the number of biological human life-years in the future accessible universe (based on current cosmological estimates) is 10^{34} years. Another estimate, which assumes that future minds will be mainly implemented in computational hardware instead of biological neuronal wetware, produces a lower bound of 10^{54} human-brain-emulation subjective life-years (or 10^{71} basic computational operations). If we make the less conservative assumption that future civilizations could eventually press close to the absolute bounds of known physics (using some as yet unimagined technology), we get radically higher estimates

of the amount of computation and memory storage that is achievable and thus of the number of years of subjective experience that could be realized.

Even if we use the most conservative of these estimates, which entirely ignores the possibility of space colonization and software minds, we find that the expected loss of an existential catastrophe is greater than the value of 10^{16} human lives.[15]

This is what connects Bostrom's work on avoiding existential risk with his work on the importance of technological and cognitive enhancement: life is valuable because it is intelligent, and a maximally intelligent life is one that is pain-free, stupidity-free, and death-free. *If* human life is worthy of existence only if it is pain-free, or at least pain-free for the most part, then it follows that—as Benatar argues—the life that we have now is not worth living. Where Benatar and Bostrom differ is not over value— both value life in its maximally capable mode, as does Singer—but in prediction: Bostrom sees human life at present as incapacitated, not yet technologically mature, and tragically subjected to a death and suffering that it ought—rationally and upon reflection—to avoid.

An extreme position, such as Benatar's, that argues for willed extinction of the human species does at least follow from his premise that only a certain type of life is worth living. We might respond to such "reasoning" that we can, and should, avoid willed extinction (of ourselves, or of a version of ourselves, or others) by shifting ethical terrain. Liberalism in its best mode would not determine in advance what counts as a life worth living, *and* would therefore go so far as to include lives that were not only *not super intelligent* but also worthy, even if not capable of the high levels of reasoning that are demanded of autonomous political subjects. As we have already seen, Martha Nussbaum has argued that we ought to include those whose lives involve different capacities and needs *and* accommodate those who must care for persons who would not meet the demands of traditional political subjects. One might even formulate a more nuanced mode of utilitarianism from such considerations: would a world in which "we" cared for those not able to care for themselves not be a more enjoyable world? Or would it not, at least, suggest values other than those of enjoyment, such as the value of experiencing human dignity, love, compassion, and care? If utilitarianism pushes us toward calculations of who ought to live, of whether life ought to be extinguished, and of weighing lives, then an expanded liberal conception of personhood would say that the very possibility of asking that question—who should live?—necessarily destroys calculus and pushes us to the question of *how one ought to live,* which in turn precludes the possibility of anyone having the expertise or measure of deciding on the being or nonbeing of other humans.

What a relief. We have done away with the awful concept of weighing of lives. We allow every person to decide what counts as being human. And for those not blessed with the power to decide, we also allow for those who must care for humans who don't quite meet the conditions of liberal personhood. Get rid of blindly rationalized utilitarianism and you get rid of the specter of extinction. Unfortunately, if some forms of rational calculus seem to foist the problem of human nonbeing before us, the problem of human nonbeing (or imminent extinction) drags us back into utilitarianism. This is very clear in more applied versions of utilitarianism and especially the discourse of health economics, where distributions and doing good can be determined by calculating QALYs (quality adjusted life-years) or DALYs (disability adjusted life-years); we might want to reject utilitarianism and health economics rationalizations, but I would suggest that the luxury of refusing calculus has always been a luxury *for some.*[16]

Tim Mulgan, in a thought experiment that writes the history of philosophy from the "broken world" of the future, argues that just as we look back with horror and puzzlement on ancient Greece and its notions of philosopher kings and natural slaves, so the future "broken world" of resource depletion will look back with wonder at the world of free liberal personhood that could proceed *without calculation* or "survival lotteries."[17] This world (of ours, today) will appear as a bizarre exception to a future world that inevitably confronts questions of who ought to survive. Not everyone can live, and not all lives are viable. If we are faced with a world of limited resources, where the life of the liberal person and favorable conditions are simply not sustainable, then however we might want to avoid it, we will be forced to ask about what counts as a viable life.

Mulgan's future broken world of survival lotteries, or a world in which some humans—because of the sheer luck of the draw—do not survive has not only already arrived: *it has always been present.* Was there ever a time when the world came even vaguely close to John Rawls's "favorable conditions," where justice was the same for all?[18] I would suggest, in a manner that differs from that of Benatar, that what has emerged *as human,* as man, is constitutively disabled, *and* that if there is anything like a sustainable life, it is precisely the life that has been extinguished in the name of the valuable and capable (or super-intelligent) human.

Rather than reject utility and calculus because of the offensive it directs to those lives it deems to be incapable, disabled, or unworthy, I would suggest that by its own calculus, the "man" of liberal reason who both generates *and* refuses utility is maximally self-disabling. By the same token, the figures of life that seem to demand nonbeing are perhaps the only forms of humanity that do not, by their own calculation, generate a calculus that leads inexorably to extinction.

As a case study, I would like to consider a case of extinction or genocide, where one group of humans decided that the human species could—possibly—benefit by eliminating one of its kind who was not quite of its kind. The use of the term *genocide,* or talking about the extinction of a race, has a recent and problematic history. One has to accept the concept of a genus of the human species to target distinct kinds of humans. A certain racial logic pertains both to targeted genocides but also in more well-meaning claims that certain events of colonial violence are best thought of as events of genocide. In the case of the "last Tasmanian Aborigine," there might seem to be some political value in identifying British colonialist strategy as a genocidal regime aiming to "breed out the color" of the Australian Aboriginal peoples.[19] Mourning the loss of a people and focusing on irrevocable loss might go some way toward forcing contemporary Australians to realizing that the past is not the past and that the drive toward the extinction of a people is not extinct. In the conclusion of this chapter, I want to question the genocidal logic that lies behind claims for lives worth living, and for human capacities that are distinct from species membership, while at the same time recognizing that the use of the term *genocide,* for all its assumptions that humans can be grouped into species and genus, is always an *offensive* (agonistic) strategy. One thing that one has to deal with, or deal with to set aside in the discussion of genocide in Australia, is the kerfuffle that became known as the history wars. If you research online about the genocide or breeding out of the Aboriginal peoples, you will come across the highly informative website of Keith Windschuttle, whose work is motivated by the desire to rid the Australian collective psyche of what he deems to be a pathological guilt and mourning.[20] One of the motivating contexts for his work was the 1997 commonwealth government report on the stolen generation, which detailed the ongoing strategy from 1910 to 1970 of removing Aboriginal children from their families.[21] One way in which this strategy was understood was as an attempt to breed out color, and it is this notion that Windschuttle rejects: what occurred may have been lamentable and part of a broad strategy of colonialism, but it was *not* genocide. If one wants to challenge Windschuttle's account, it would make sense to emphasize race and not to say that deep down we are all human and therefore what took place was "merely" colonization. If one does not recognize race, then one is blind to racial strategy, and if one does not recognize genocidal strategy, then one does not recognize the ongoing specter of a particular type of assimilationist violence. However, one further problem attends the strategy of claiming that genocidal intent was directed against Australian Aboriginal peoples: the mourning (by way of a highly languid 1978 television documentary) of the last Tasmanian not only displaces colonial violence to a different time and place but also maintains the notion of "a" race that could be isolated and extinguished and implicitly claims that there are now no persons who might claim land rights on the basis of being tied to the land.[22] The "extinction" of "the" Tasmanian Aborigine is at once a cultural fantasy about a violent colonization that is well and truly in the past *and* an erasure of other modalities of being human that "we" mourn as lost. On one hand, nonindigenous Australians *need* the notion of Aboriginals who are tied to land by way of a timeless dreaming rather than ownership or filiation: there must be, somewhere, a sense of space and time that is not that of managerial capitalism. And yet, it is precisely that thought of *another* humanity—one that was sufficiently other to the point that it could be extinguished—that allowed claims that Tasmanian

Aborigines were extinct (and therefore no longer a burden for land rights claims). Once again, it is a certain type or form of subject that can look at the array of human lives and claim that "a" race has become extinct; this purveying eye that has a command of history, anthropology, life, and time both requires and erases any mode of "the human" other than its own capable kind. We seem to be poised, as liberal multiculturalism often is, between postracial claims for a general humanity that does not need to be marked or set apart to achieve a right and a politically astute account of the ways in which white colonizing capitalism achieved its universality by erasing and exterminating others, and creating them *as other* by way of strategies of cultural erasure. But I want to suggest that this seemingly intractable and universal problem is a problem for a portion of humanity, and a portion that has the logic of extinction at its heart.

Let us go back to the first problem of who ought to live and why, or the question of how one ought to live, and what counts as a good life or a life worth living. As I suggested, problems of extinction bring in, it seems, a form of utilitarianism: how do we manage the survival of life, maximizing life, and maximizing *good life*? At the same time, questions of utility seem to raise the specter of extinction: some lives might just not be worth living. But perhaps these questions are already racial, bound up with the "man" of Western reason who is *not* a species. From the Socratic elevation of the examined life to the various forms of posthumanism that range from assuming that there is a prima facie value attached to the ongoing survival of thinking to the inclusion of nonhumans as persons, white Western man does not have a race: but he does not have a race because he asks the question of the value of life, of *what it means to live. He* is at once the only man to face extinction—for when we view contemporary cinema and television about the end of the world, it is *this* man (the man of libraries, the familial man, the postracial man, the man of reason) who is threatened with extinction, or the world's end. What we witness is not genocide but the end of the *world*. It is because this man has always asked about "the good life"—even if that is a liberal life that has no good other than the asking of the question—that he can be the victim of extinction. Asking the question of the good life, of how "one" ought to live, is both genocidal and extinction generating.

Since its invasion, Australia was deemed to be *terra nullius* partly on the basis of rampant opportunism but also because a form of (indigenous) life was not recognized as properly human. Not only were indigenous Australians not property owning, industrious, and industrializing developers of the land with a technoscience oriented to the maximization of a life they identified as human, and thereby not deemed worthy of recognition, but the very logic of technoscience that could only recognize such cultures as minor and racial (distinct, enigmatic) would be the same planet-transforming "species" that now proclaims itself as author of the Anthropocene. The very possibility of utilitarian questions—who ought to live? is this a life worth living? how might we live on *maximally?*—is part of a logic of appropriation, extension, survival, and calculus that divides species/genus questions. There are metaphysical questions—about how "one" ought to live and the life worth living—and these are for *man,* who is not a species but a potentiality, a power of thinking and living that transcends any body. And then there are genus questions: how "we" negotiate different claims for survival. It makes sense to mourn the extinction of "the" Aboriginal people, for *those* people have a race that might survive only by way of blood, language, culture, and a distinct archive.

To conclude, I would note first that what counts as the individual of ability—where self and ability are mutually constitutive—is at the heart of the Anthropos who has precipitated itself and others into accelerated extinction. The self of technoscience can easily be tied to the pollution of Earth, but so can the universalizing self of liberal and utilitarian theory: I can kill, exterminate, and save *if* I have the ability to think beyond myself to the curious value of life as such, of life that might be maximized and weighed.

The self of disability might appear to be of secondary or parasitic concern, but I would argue the contrary: it is organic disability that requires a body to generate *techne,* stored energy and archives; the more this dependence is mastered, the more a disequilibrium opens up between those who render themselves productively and theoretically able and those who possess different abilities and disabilities. I am not just saying that had "we" not developed all those abilities that are definitive of the liberal subject, the Earth would be better off, as though human excellence came at a price to nonhumans; I am saying that the very questions of how one ought to live, of the value and meaning of life, of weighing life, create a specific terrain and orientation that is now reaching its limit.

It is not, then, that the self of liberalism and utilitarianism needs to expand and include other modes of the self, to be more caring to those *not* blessed with the same abilities; that self needs to be seen not as the basis of the species that must be saved but as a genus tied inextricably to logics of extinction. One can only calculate the worth of living, at the expense or cost of other lives, if one has a conception of *life,* and it is that general conception that is both historically and culturally odd, and that requires an anthropology. How did some living beings constitute themselves as bearing an ability to evaluate life? How is it possible for a being to ask about the value of one life as opposed to another life? How is it that the agonistics of life became a calculus? I want to point out not only that there are many modes of being human for whom the overall existence and extinction of "the human" is not a problem but that the modality for whom extinction of intellectual life *is a problem* is a self of white, modern, calculative ability that is exceptional and *not* the default setting of the species—if there is such a thing.

Gilles Deleuze, writing on Foucault, points toward the specificity of the man-form that comes into being by way of a certain type of question.[23] What allows something like "man" to emerge is that, rather than seeing his being as an aspect of a complex whole that he knows with some degree of clarity and distinction, he comes to know himself clearly and distinctly, and then places what is other than himself—nature, life, the biological or species being of the human—in parentheses. We are distanced from that life, but that distance or absence of foundation allows us to become self-legislating, contractual, formally rational subjects. Life does not tell us what to do, and we are not simple expressions of life; the human, or man *as question,* must now labor over whether all life makes a claim to be, or whether the being who asks that question—a being liberated from mere life—has some privilege: do we save the local, indigenous, immediate, and unreflective, or does the capacity to ask that question create every other form of life as one expression of anthropological calculating man? When philosophers dispute about a life worth living, arguing whether a life is able enough to live, they are part of the same voice that can observe fragments of the human species as a genus, or a particular kind of a general species, over which a single voice might range. End-of-world narratives, and scenarios of catastrophic risk—such as those of Nick Bostrom— contemplate the extinction of this "genus which is not one" and assume both that *this* would be the catastrophe of all catastrophes and that humanity is necessarily defined by a certain concept of personhood that is irreducible to the human species. Indeed, it is ability—in Bostrom's case, intelligence—that needs to be preserved; it is *this* life that would count as extinction as such and not "merely" as genocide. An anthropological and calculative "we" emerges by way of technologies that generate and calculate the worth of "a" life, and this life is the life of a person: a being who is distinct from nature, and who may even calculate something like her own right to life or cost to the earth by way of a carbon footprint, imagining that she might live on this earth but deftly erase any damage to her milieu. It is this same person, distinct by way of certain predicates, who might view and weigh other human nonpersons as members of a genus, as instances of a way of life to be preserved, *or not.*

NOTES

1. Martha C. Nussbaum, *Frontiers of Justice: Disability, Nationality, Species Membership* (Cambridge, Mass.: Harvard University Press, 2006).

2. Bernard Stiegler, *Technics and Time: The Fault of Epimetheus,* trans. Richard Beardsworth and George Collins (Stanford, Calif.: Stanford University Press, 1998), 122.

3. Nick Bostrom, "Existential Risk Prevention as Global Priority," *Global Policy* 4, no. 1 (2013): 15–31.

4. Nussbaum, *Frontiers of Justice,* 160.

5. Ibid., 129.

6. Lennard J. Davis, "Introduction: Disability, Normality and Power," in *The Disability Studies Reader,* ed. Lennard J. Davis (New York: Routledge, 2013), 1–17; Giorgio Agamben, *The Use of Bodies,* trans. Adam Kotsko (Stanford, Calif.: Stanford University Press, 2015).

7. Peter Singer, *Practical Ethics* (Cambridge: Cambridge University Press, 1993).

8. Timothy W. Luke, "On the Politics of the Anthropocene," *Telos* 172 (Fall 2015): 139–62, doi:10.3817/0915172139.

9. Rebecca Newberger Goldstein, "How Philosophy Makes Progress," *The Chronicle of Higher Education,* April 14, 2014, http://www.chronicle.com/article /Is-Philosophy-Obsolete-/145837/.

10. David Benatar, *Better Never to Have Been: The Harm of Coming into Existence* (Oxford: Clarendon Press, 2006).

11. Singer, *Practical Ethics,* 182.

12. Peter Singer, *Rethinking Life and Death* (London: Macmillan, 1996), 213.

13. Ibid., 216.

14. Derek Parfit, *Reasons and Persons* (Oxford: Oxford University Press, 1984), 75.

15. Bostrom, "Existential Risk Prevention," 18.

16. Christopher J. L. Murray, "Rethinking DALYs," in *The Global Burden of Disease,* ed. Christopher J. L. Murray and Alan D. Lopez, 1–98 (Cambridge, Mass.: Harvard School of Public Health on behalf of the World Health Organization and the World Bank, 1996).

17. Tim Mulgan, *Ethics for a Broken World: Imagining Philosophy after Catastrophe* (Montreal: McGill-Queen's University Press, 2011).

18. John Rawls, *Theory of Justice* (Oxford: Clarendon Press, 1972).

19. Russell McGregor, "'Breed out the Colour' or the Importance of Being White," *Australian Historical Studies* 33, no. 120 (2002): 286–302.

20. Keith Windschuttle, "Why There Were No Stolen Generations (Part One)," *Quadrant Online,* January 1, 2010, https://quadrant.org.au/magazine /2010/01-02/why-there-were-no-stolen-generations/.

21. Commonwealth of Australia, *Bringing Them Home: Report of the National Inquiry into the Separation of Aboriginal and Torres Strait Islander Children from Their Families* (Sydney: Commonwealth of Australia, 1997), https://www .humanrights.gov.au/sites/default/files/content/pdf/social_justice/bringing _them_home_report.pdf.

22. Anne Curthoys, "Genocide in Tasmania: The History of an Idea," in *Empire, Colony, Genocide: Conquest, Occupation and Subaltern Resistance in World History,* ed. A. D. Moses, 229–52 (New York: Berghahn Books, 2008).

23. Gilles Deleuze, *Foucault,* trans. Seán Hand (Minneapolis: University of Minnesota Press, 1988).

8

Biocapitalism and De-extinction

Ashley Dawson

A snow leopard roamed down the face of the Empire State Building. A beautiful, charismatic endangered animal, it was followed by a dazzling Kaiser's spotted newt, a type of salamander indigenous to Iran. The animals were projected onto the side of New York City's iconic skyscraper by two artists during a week in 2015 in which the plight of the world's wild animals was highlighted in a number of ways.[1] Traveling in his father's natal country of Kenya, President Obama announced restrictions on the sale of African elephant ivory in a bid to contain the heightening decimation of the continent's population of pachyderms. That same week, an American hunter shot and killed a lion named Cecil in Zimbabwe, sparking a scandal and bringing international attention to the endurance of big game hunting, a "sport" many had thought died with the colonial era. While hunters offered justification for their practices by arguing that the fees they pay help sustain populations of "wild" animals, who are carefully reared before being culled by well-heeled dentists and other tourists to underdeveloped nations, the underlying structures of feeling that once sustained the hunting of "game" animals have shifted dramatically since the colonial era. For characters in Victorian colonial novels such as H. Rider Haggard's *King Solomon's Mines,* which is set in the country where Cecil the lion was killed, hunting was an affirmation of the valorous masculinity of the white male colonist, an act that linked him to his (putative) chivalric forebears, who earned their aristocratic titles—and the land that came with them—through acts of martial prowess on the battlefield and in the hunt. Armed with rifles and shotguns, colonial hunters massacred a significant portion of Africa's wildlife before realizing that they needed to engage in "conservation." A sea change in attitudes toward wild animals took place, and preserves were established in many postcolonial countries around the world to protect them. Public opinion in Western nations is currently undergoing a similar transformation, as understanding spreads that we are in the midst of the sixth great mass extinction in the planet's history, an event that forms the backdrop to the endangered animals crawling and romping across the Empire State Building.

What is to be done in response to the sixth extinction? The answer to this question depends on how the causes of the crisis are framed. All too often, the roots of extinction are divined in some general capacity of human beings to destroy the natural world. Near the conclusion of her exploration of the sixth extinction, for example, science journalist Elizabeth Kolbert argues that extinction happens because the world's flora and fauna cannot adapt to the accelerated rate of change human beings are imposing on the world. "As soon as humans started using signs and symbols to represent the natural world," she writes, "they pushed beyond the limits of that world."[2] The very qualities that make us human—our creativity, our communicative and collaborative abilities—have, according to Kolbert, allowed us to transform the world in ways that empower us but also endanger the natural world on which we ultimately depend. But note the sweeping universalism in Kolbert's account of the forces driving extinction. Humanity is represented as unified and undifferentiated, as if we are all equally culpable for the current wave of extinction. Indeed, later in that same paragraph, Kolbert writes, "If you want to think about why humans are so dangerous to other species, you can picture a poacher in Africa carrying an AK-47 or a logger in the Amazon gripping an ax, or, better still, you can picture yourself, holding a book on your lap."[3] Kolbert seems to suggest that the forces driving extinction are propelled not just by rifle-packing game hunters but also by environmental enthusiasts whose concern about endangered wildlife might have led them to purchase her book. But this sweeping indictment not only spreads blame around to the point that everyone and no one may be held accountable. It also begins with the image of an African poacher and a logger in the Amazon, who presumably are the true perpetrators of habitat destruction. No mention is made of the transnational logging companies, agribusiness interests, or mineral-extraction conglomerates at whose behest these figures usually work.

The scapegoating of indigenous peoples in the Global South as the culprits for extinction evident in Kolbert's account emerges from a long history of flawed conservation efforts. Faced with fresh concerns about an escalating wave of extinction, it is worth recalling that conservation since the end of the colonial era has had highly destructive effects on indigenous peoples who live in the areas established for protection. In his book *Conservation Refugees,* Mark Dowie quotes delegates to the Fifth World Parks Conference who say, "We were dispossessed in the name of kings and emperors, later in the name of state development, and now in the name of conservation."[4] All too often, indigenous people who lived among the animals once massacred by colonial hunters were not only ignored in the process of conservation, rendered invisible in the campaign to preserve valuable game animals, but were regarded as a threat to land and wildlife conservation. The scapegoating of people in the name of conservation continues today. Residents of protected areas are often labeled with what Rosaleen Duffy calls "nature crime" and subjected to harsh punitive measures in the name of protecting land and wildlife.[5] Over the last century, efforts to preserve endangered land and wildlife have led to the creation of millions of "conservation refugees." Reacting to this history and to present instances of persecution, two hundred delegates to a meeting of indigenous peoples signed a declaration stating that the "activities of conservation organizations now represent the single biggest threat to the integrity of indigenous lands," eclipsing the long-term threat of extractive industries.[6] But, belying the opposition between extraction and conservation in this statement, international conservation organizations are often quite willing to work hand in glove with global resource prospectors, who fund their efforts handsomely.[7]

In light of this history of persecution, it is imperative that extinction be viewed through a more critical, materialist lens.[8] The sixth extinction needs to be framed not as a product of some general human capacity for despoliation of the planet—as we see in the work of journalists like Kolbert—but rather as the product of a global attack on the commons, a capitalist frenzy as the planet tilts toward increasingly intense environmental catastrophe. The destruction of global biodiversity should be seen, in other words, as a great, and perhaps ultimate, attack on the planet's commonwealth, one in which capitalist interests target the world's remaining stores of biodiversity. Though this attack is playing out largely—but certainly not exclusively—in the Global South, it emanates from the economically and politically dominant nations in the Global North. Indeed, extinction needs to be seen, along with climate change, as the leading edge of contemporary capitalism's contradictions.[9] If capital must expand at an ever-increasing rate or go into crisis, "development" is now consuming entire ecosystems and thereby threatening the planetary environment as a whole. The catastrophic rate of extinction today and the broader decline of biodiversity thus represent a direct threat to the reproduction of capital. Indeed, there is no clearer example of the tendency of capital accumulation to destroy its own conditions of reproduction than the sixth extinction. Nature, the wonderfully abundant and diverse wildlife of the world, is essentially a free pool of goods and labor that capital can draw on. As the rate of speciation—the evolution of new species—drops further and further behind the rate of extinction, the specter of capital's depletion, and even the annihilation of the biological foundation on which it depends, becomes increasingly apparent.

In what follows, I explore the ways in which the extinction crisis offers an opportunity to capital for a new round of accumulation. In the name of coping with the decimation of flora and fauna around the planet, the most advanced sectors of capital are rolling out new biotechnologies that, we are told, promise to revive charismatic extinct species like the mastodon. These new techniques of genetic engineering confer godlike capacities not simply to restore extinct species but also to rewire life to conform to the dictates of corporate profit. In the second section of the chapter, I explore the ideology of ecomodernism that plays an important role in legitimating the new biocapitalism. Ecomodernism, I argue, is predicated on facile ideas of increasingly beneficial economic development that collapse under scrutiny. Finally, the concluding portion of the chapter examines one particular geographical site—the Amazon—where ecomodernist ideas about development and the need to increase the intensity of energy production are being deployed, with disastrous effects on biodiversity and indigenous culture. Plans for a massive regionwide development scheme in the Amazon provide the most palpable and alarming possible example of the willingness of contemporary capitalist culture to engage in a fresh bout of what Joseph Schumpeter called "creative destruction" in the face of the sixth extinction.[10] Yet it is worth remembering that, in contrast with his neoliberal acolytes, Schumpeter followed Marx in arguing that this tendency of capitalism would ultimately destroy the material foundation on which it rested.[11] This bout of cataclysmic destruction targets the peoples and other life-forms of the Global South in a new wave of imperialism. The sixth extinction thus calls for the creation of fresh forms of solidarity and anti-imperialism based on novel understandings of the interwoven character of all life on Earth.[12]

DE-EXTINCTION AND REGENESIS

De-extinction, the most seemingly radical solution currently proposed for the extinction crisis, promises to wind evolutionary time backward. Efforts to use traditional methods of back-breeding to restore an approximation of extinct species have been around since the Heck brothers attempted to revive the aurochs, the ancestor of all the world's cows, in Germany during the 1920s. Today, the Tauros Program is attempting to re-create the aurochs using similar back-breeding techniques, now guided by specific knowledge about the aurochs genome drawn from molecular biology and genetics. The goal is to release the Tauros, a bovine breed that proponents claim will be indistinguishable from the aurochs, into European rewilded areas such as Holland's Oostvaardersplassen by 2020.[13] But such efforts are conservative in comparison to the goals of de-extinction advocates who embrace the full potential of genomic technologies to resurrect extinct species like the wooly mammoth. An extinct animal has already been brought back to life: in 2000, a Franco-Spanish team transferred the nucleus of a skin cell from the world's last Pyrenean ibex, which was found dead in northern Spain earlier that year, into the egg cell of a domestic goat, implanting the cell into a surrogate mother in a process called interspecies nuclear transfer cloning.[14] Although the baby ibex died shortly after birth, the experiment showed that an extinct species could be brought back to life. For scientists like George Church, inventor of MAGE (multiplex automated genomic engineering) technology, synthetic biology promises nothing short of the resurrection of any extinct species whose genome is known or can be reconstructed from fossil remains.

Key to this process is the conceptualization of animal species as bundles of genetic information, sequences of letters that can be stored on a computer. Animals (and humans, for that matter) are nothing more than genetic code, in this view, easily transposed into computer code.[15] Once the genomic information of a particular species is recovered or decoded, the problem simply becomes converting that information into strings of nucleotides that constitute the genes and genomes of the animal.[16] Church's MAGE technology massively accelerates the techniques of genetic engineering, allowing scientists to take the intact genome of one animal, an elephant, say, and rejigger it using another animal's genome, a mammoth's, for example, as a template, thereby creating a new genome that can be used to clone a live mammoth into existence using the same transfer cloning process used to bring back the extinct Spanish ibex. Church, with the aid of organizations such as Revive and Restore, hopes to resurrect not just the mammoth but also the passenger pigeon, the Caribbean monk seal, the golden lion tamarin, and many other now-extinct animal species.[17]

Synthetic biology promises, in Church's words, "to permit us to replay scenes from our evolutionary past and to take evolution to places where it has never gone."[18] De-extinction is thus intimately tied to regenesis. MAGE has in fact been nicknamed "the evolution machine," because it can carry out the equivalent of millions of years of genetic mutations within minutes. The biblical resonance in the notion of regenesis is not ironic. Synthetic biology, proponents argue, will quite literally make us into gods by allowing us not just to resurrect extinct life-forms but also to engineer new life according to our needs.[19] Not surprisingly, this promise has fired the public imagination and attracted a fair amount of venture capital to organizations like Revive and Restore. But even the most ardent proponents of de-extinction admit that many problems remain to be resolved. Biologists are, for instance, ironically far better at manipulating genomes than at rewilding landscapes.[20] Furthermore, does it really make sense, many conservation biologists wonder, to spend huge amounts of time and capital to resurrect an extinct species such as the passenger pigeon when the threats that were responsible for its extinction—such as habitat destruction—have only intensified? We may be able to resurrect a few individual specimens of an extinct species as a spectacle, but will we not cruelly doom them to become extinct once again if we place them back in denuded habitats?

De-extinction offers a seductive but dangerously deluding techno-fix for an environmental crisis generated by the systemic contradictions of capitalism. It is not simply that de-extinction draws attention—and economic resources—away from other efforts to conserve biodiversity as it currently exists.[21] The fundamental problem with de-extinction is that it relies on the thoroughgoing manipulation and commodification of nature and, as such, dovetails perfectly with biocapitalism. U.S. lawyers have already begun arguing that revived species such as the mammoth would be "products of human ingenuity" and should therefore be eligible for patenting.[22] Species revival thus slots seamlessly into the neoliberal paradigms of research established by the Bayh-Dole Act of 1980, which made the patenting of scientific inquiry legal, as well as with the intellectual property agreements foisted on the world since the establishment of the World Trade Organization in the mid-1990s.[23] De-extinction provides a mouthwatering opportunity for a new round of capital accumulation based on generating, and acquiring intellectual property rights over, living organisms. It is perhaps the most tangible and fully realized example of a shift that has been taking place since the 1980s, in which U.S. petro-chemical and pharmaceutical industries have reinvented themselves as purveyors of new, clean life sciences. Instead of generating (declining) profits through the mass-produced chemical fertilizers and pesticides of the Fordist era, agribusiness corporations like Monsanto have repositioned themselves to generate life itself by buying up biotech start-up companies. Capital is shifting, as Melinda Cooper observes, into "a new space of production—molecular biology—and into a new regime of accumulation, one that relies on financial investment to a much greater extent."[24] In this new postmechanical age of production, the biological patent allows a company to own an organism's principle of generation, its genetic code, rather than owning the organism itself. Biological production is thereby transformed into capital's primary means for generating surplus value. Under this new regime of biocapitalism, living organisms are increasingly viewed, in the words of George Church, as "programmable manufacturing systems."[25]

Biocapitalism is generated by and deeply embedded in U.S. imperialism. The massive investments in the life sciences that characterize this regime of accumulation are a product of the monetarist counterrevolution of 1979–82, when the United States introduced interest rate policies that channeled global financial flows into the dollar and U.S. markets.[26] Since then, the United States has financed its perpetually spiraling budget deficits through continuous inflows of capital. The result has been a form of capitalist delirium, which enables the United States to operate—for a time—in utter disregard of economic and ecological limits. Yet U.S. debt imperialism is based on the extraction of capital from vassal nations through the imposition of crippling structural adjustment policies by organizations like the International Monetary Fund and the World Bank.[27] Prostrated by debt, developing nations have been forced to sell off public assets and to open their economies to external capital penetration in a series of global enclosures of the commons. Ignoring these conditions of accumulation by dispossession, however, the ideologues of biocapitalism draw on the work of scientists such as Ilya Prigogine, whose *Order Out of Chaos* challenged the notions of limits inherent in the second law of thermodynamics by arguing that all of nature obeys the laws of self-organization and increasing complexity that characterize biological processes and systems.[28] Like life itself, neoliberals under the sway of this biocapitalist paradigm came to argue, the economy is characterized by a process of continuous, self-regulating *autopoiesis* or self-engendering.[29] And, again like life, capitalism is said to be characterized by a series of catastrophic crises that ultimately generate new forms of complexity, as do mass extinction events in evolutionary history.

These neoliberal ideologies have come to permeate conservation to such an extent that discussions of biodiversity have become the site for the elaboration of *disaster biocapitalism.* Just as the disaster capitalism described by Naomi Klein seizes on political calamities to further its accumulative aims, this disaster biocapitalism takes the extinction crisis as an opportunity to ratchet up the commodification of life itself.[30] At the UN Climate Conference in 2007, for example, the United Nations (UN) and the World Bank announced the Reducing Emissions from Deforestation and Forest Degradation (REDD) scheme, which pays countries of the Global South to reduce their deforestation and protect their existing forests. The carbon stored in these forests can then be quantified and sold to polluting industries in the Global North, who can buy this stored carbon to "offset" rather than reduce their own polluting emissions. REDD was launched without input from indigenous peoples and other forest-dependent communities and has already been linked to many land grabs and human rights violations.[31] All too often, local land stewards are represented in corporation-controlled international agreements like REDD as destroyers of biodiversity and are consequently subjected to forced removal so that ecosystems can be privatized and reengineered as income-generating commodities to be sold on global capital markets.

ECOMODERNISM, OR HOW I LEARNED TO STOP WORRYING AND LOVE GREEN CAPITALISM

One of the prime movers of the de-extinction Revive and Restore initiative, president of the Long Now Foundation, and one-time editor of the 1960s countercultural bible known as the *Whole Earth Catalogue,* Stewart Brand is also a key player in the controversial ecomodernism movement. Proponents of ecomodernism challenge the established environmental movement in the *Ecomodernist Manifesto,* a tract coauthored by Brand and seventeen other researchers associated with the Oakland-based Breakthrough Institute.[32] Founded in 2003 by pollster Ted Nordhaus and lobbyist Michael Shellenberger, the Breakthrough Institute has issued stinging criticisms of the preservationist tenets of mainstream environmentalism based on Nordhaus and Shellenberger's credo that "the solution to the unintended consequences of modernity is, and always has been, more modernity."[33] True to this tradition, the *Ecomodernist Manifesto,* appropriating the now-ubiquitous concept of the Anthropocene—the geologic age during which human beings have significantly marked the planetary environment—argues that "a good Anthropocene demands that humans use their growing social, economic, and technological powers to make life better for people, stabilize the climate, and protect the natural world."[34] For Brand and other ecomodernists, industrialization, technological innovation, and economic development are not only compatible with ecological sustainability but key drivers of environmental reform. Rather than opposing the reckless growth imperative that animates contemporary capitalism, in other words, ecomodernists believe that unfettered capitalism will save the planet. This dovetails nicely with the biocapitalist imperatives of de-extinction.

Key to the ecomodernist embrace of various forms of biocapitalism is the concept of *decoupling*. As the *Ecomodernist Manifesto* puts it, "intensifying many human activities—particularly farming, energy extraction, forestry, and settlement—so that they use less land and interfere less with the natural world is the key to decoupling human development from environmental impacts" (7). What precisely do the ecomodernists mean by decoupling? Ecomodernists believe that certain cultural forms and technological developments hold the potential to diminish human beings' impact on the environment. Among these are the following: urbanization, agricultural intensification, nuclear power, aquaculture, and desalination (18). This catalog bears a strong resemblance to the technologies celebrated in Brand's recent book *Whole Earth Discipline,* whose subtitle itemizes his "ecopragmatist" embrace of dense cities, nuclear power, transgenic crops, restored wildlands, and geoengineering.[35] These technologies, Brand and his fellow ecomodernists argue, hold the potential to alleviate the impact of human development on the natural world. To support this argument, the ecomodernists distinguish between relative and absolute decoupling. According to the manifesto, "relative decoupling means that human environmental impacts rise at a slower rate than overall economic growth. Thus, for each unit of economic output, less environmental impact (e.g., deforestation, defaunation, pollution) results. Overall impacts may still increase, just at a slower rate than would otherwise be the case. Absolute decoupling occurs when total environmental impacts—impacts in the aggregate—peak and begin to decline, even as the economy continues to grow" (11). The ecomodernists point to the peaking of the population growth rate of human beings, to the urbaniza-tion of humanity, and to increasing agricultural intensification as trends that, taken together, suggest that "demand for many material goods may be saturating as societies grow wealthier" (14).

The ecomodernists are not complete Pollyannas. They do admit that, while "nations have been slowly decarbonizing—that is, reducing the carbon intensity of their economies"—since the Industrial Revolution, they have "not been doing so at a rate consistent with keeping cumulative carbon emissions low enough to reliably stay below the international target of less than two degrees centigrade of global warming" (20). In addition, in a clear nod to Brand's work with Revive and Restore, they allude to the global extinction crisis, writing that "human flourishing has taken a serious toll on natural, nonhuman environments and wildlife. Humans use about half of the planet's ice-free land, mostly for pasture, crops, and production forestry. Of the land once covered by forests, 20 percent has been converted to human use. Populations of many mammals, amphib-ians, and birds have declined by more than 50 percent in the past forty years alone" (9). Rather than concluding from these sobering facts that the capitalist economic system has put humanity on a path to ecocide, the ecomodernists instead argue that an intensification of capitalist culture, including a ramping up of global energy production, is necessary to bring about "climate stabilization" and "radical decoupling." It is worth noting that none of the technologies they propose as "plausible pathways" to these ends—next-generation solar, advanced nuclear fission, nuclear fusion—currently exist. During the "transition" period before such tech-nologies are developed, the ecomodernists argue that "hydroelectric dams may be a cheap source of low-carbon power for poor nations even though their land and water footprint is relatively large" (24). As we will see, such prescriptions ignore not simply the environmental impact of big dams in the Global South but also discount decades of intense struggle waged against such technologies by indigenous and poor people's movements.

Brand and his collaborators characterize themselves as heretics bring-ing dissident and fresh perspectives to a hidebound environmental move-ment. Despite this contrarian stance, their arguments are completely consistent with the central myth of global capitalist culture: the belief in the possibility of unending growth. For the entire span of the "American Century," growth has been the most important policy goal of governments of all stripes around the world. The global economy is, in fact, five times the size it was after World War II and is expanding at such an exponential rate that it will be eighty times that size by 2100. As commentators such as David Harvey have pointed out, this rate of growth is totally at odds with the finite resource base and increasingly fragile environment on which humanity depends.[36] To obscure the glaring contradiction of infinite economic expansion on a self-evidently limited planet, advocates of growth typically allude to the myth of decoupling. The argument goes that because capitalist economies are inherently efficient, the capitalist system as a whole can continue to expand while requiring continually declining material throughput. Such arguments constitute a form of conventional thought, one might even say of dogma, among mainstream economists. The ecomodernists are thus deeply complicit with the dominant beliefs of our turbocapitalist age. But are their accounts of decoupling correct?

The answer to this question hinges on distinguishing between relative and absolute decoupling. There is significant evidence for declining resource intensities, or relative decoupling. Over the last thirty years, for example, global carbon intensity fell from around 1 kilogram per dollar of economic activity to just under 770 grams per dollar.[37] But improvements such as these in carbon and energy intensity were more than offset by massive increases in the scale of economic activity in general. Not only is there therefore no evidence for overall reductions in resource throughput (absolute decoupling) but global carbon emissions from energy use have actually *increased* by 40 percent since 1990, the Kyoto base year.[38] In addition, when one factors in global consumption of a range of nonfuel minerals such as iron ore and bauxite, global resource intensities have actually intensified dramatically in recent decades. As Tim Jackson explains, estimated rates of declining carbon intensity roughly equal projected population growth rates, but under business-as-usual scenarios, annual growth rates of 1.4 percent in income mean that by 2050, the advanced industrialized nations will be emitting *80 percent* more carbon than we are at present.[39] As these statistics underline, the ecomodernists' assertions about humanity's economic and technological prowess as a solution to the environmental crisis is predicated on a failure to distinguish adequately between relative and absolute decoupling. One need simply look at the relentless increase of carbon emissions to see the folly of their fetishization of technological progress. On a more fundamental level, their arguments that capitalism's tendency toward greater efficiency will permit us to stabilize the climate and avert looming environmental catastrophe are based on totally delusional beliefs that prop up a deeply unjust global status quo.

The ecomodernists should be seen in relation to the specific cultural formation that shaped many of them, a belief system that has been dubbed the "California Ideology."[40] Born of the countercultural hippy movement, the California Ideology has been thoroughly assimilated by the highly capitalized hi-tech firms of Silicon Valley; nevertheless, the California Ideology is a structure of feeling that allows its proponents to believe they are being contrarian while espousing now-dominant forms of asocial libertarianism. Equally central to this structure of feeling is a deification of technology that allows proponents of this ideology to believe steadfastly in the emancipatory potential of information technologies. As the cultural and economic impact of digital culture has spread in recent decades, the California Ideology has diffused across the United States. The U.S. environmental movement is increasingly saturated with such pro-capitalist, neoliberal ideology, from Newt Gingrich's paeans to technologically savvy environmental entrepreneurs to Paul Hawken and Amory Lovins's boosterism for "natural capitalism."[41]

The extent of the ecomodernists' embrace of current trends can be quite shocking. In a recent article on the question of extinction, for instance, ecomodernist and de-extinction guru Stewart Brand argues that climate change is likely to be a boon rather than a death sentence for the world's flora and fauna. Quoting the work of conservation biologist Chris Thomas, Brand argues that "what we might be seeing in response to climate change . . . 'is starting to look very much like a global acceleration of evolutionary rates.'"[42] It's hard to believe that Brand really thinks that the great mass of life on the planet at present will be capable of evolving to cope with the massive warming, ocean acidification, and other environmental holocausts fossil capitalism is currently catalyzing. Some life will no doubt survive, and biodiversity may rebound after millions of years, as Brand argues, but there will be great decimation in the interim. But it's not just a question of temporal scale. Sometimes Brand simply engages in convenient misquotation. For instance, to support his argument that we are not in the midst of a human-caused sixth mass extinction, Brand cites a study from *Nature* magazine that expressed uncertainty about the exact scale and pace of extinction given our partial knowledge of existing species. Brand quotes the article as such: "If all currently threatened species were to go extinct in a few centuries and that rate continued, the sixth mass extinction could come in a couple of centuries or a few millennia." For Brand, "the range of dates in that statement reflects profound uncertainty about the current rate of extinction."[43] Perhaps, but this is a misquotation of the article, which actually reads, "If all currently threatened species were to go extinct in a few centuries and that rate continued, the die-offs would soon reach the level of a mass extinction—the kind of biological catastrophe that ended the reign of the dinosaurs and that has happened only five times in Earth's history."[44] While it's possible that Brand conveniently misread that sentence, the amputation suggests willful dissimulation. The sentence as it was written is clearly suggesting that we are in the incipient phases of an extinction crisis equivalent to that of the Cretaceous event, when half of the world's species were obliterated. Moreover, the rest of the article unequivocally contradicts Brand's flat denial of the extinction crisis. The three sentences before his misquotation, for instance, read thus: "Before human populations swelled to the point at which we could denude whole forests and wipe out entire animal populations, extinction rates were at least ten times lower. And the future does not look any brighter. Climate change and the spread of invasive species (often facilitated by humans) will drive extinction rates only higher. The pace of extinction is leading towards a crisis."[45] One couldn't imagine a much more forthright clarion call concerning extinction coming from the mouth of a scientist than this, one that pins the extinction crisis on human beings to boot.

Yet if he's skeptical about the idea of a sixth extinction, Brand does acknowledge the reality of *defaunation*—with massive declines in the numbers of keystone animal species leading to the crash of entire ecosystems. The solution to defaunation for Brand is more aggressive human intervention in the natural world. Among the measures he champions are forms of "ecological replacement," in which wild animals are translocated from their current ranges to occupy lands from which they were extirpated or to replace similar extinct species. A natural extension of this process of ecological replacement, Brand argues, is de-extinction. The biocapitalist emphasis to Brand's argument is not even thinly veiled: "In parallel with the arrival of 'precision medicine' for humans, where treatment can be specific to the genomes of individual patients (and even individual tumors), we might see the development of 'precision conservation' techniques based on minimalist tweaking of wildlife gene pools. Some conservation scientists refer to it as 'facilitated adaptation' and see it as a form of 'applied evolutionary biology.'"[46] Governments and social movements have barely begun to debate the ethical and practical implications of "minimal tweaking" of wildlife gene pools, but already there are arguments for the destruction of "nuisance species," such as the mosquito, through genetic engineering.[47] Brand's arguments make the parallel between such dubious interventions and the highly remunerative nascent field of "precision" (or designer) medicine for the hyperaffluent starkly evident.

How do the ecomodernists' biocapitalist views shape their perception of the people who inhabit endangered ecosystems? The assumptions about the beneficial character of economic growth and capitalism in general that typify ecomodernism lead the movement's advocates into some highly problematic positions regarding what Ramachandra Guha and Juan Martinez-Alier call *ecosystem people*.[48] Rather than aligning themselves in solidarity with such communities against what Guha and Martinez-Alier call the *omnivores*—local and global elites capable of capturing and using natural resources—the ecomodernists hold poor communities equally culpable for ecological crises as transnational corporations: "Ecosystems around the world are threatened today," the ecomodernists contend, "because people over-rely on them: people who depend on firewood and charcoal for fuel cut down and degrade forests; people who eat bush meat for food hunt mammal species to local extirpation. Whether it's a local indigenous community or a foreign corporation that benefits, it is the continued dependence of humans on natural environments that is the problem for the conservation of nature."[49] The ecomodernist position is remarkably naive about the history of human rights abuses meted out around the world during the last half-century in the name of development.[50] Nor do they evince any sense of the genocidal violence that subtended the consolidation of European modernity.

ECOCIDE IN THE AMAZON

In policy terms within a domestic U.S. context, ecomodernist techno-boosterism translates into opposition to efforts to limit greenhouse gas emissions, the presumption being that investment in nuclear reactors and shale gas extraction will produce all the clean energy we need.[51] But what does their manifesto-defining assertion that "plentiful access to modern energy is an essential prerequisite for human development and for decoupling development from nature" imply in the developing world?[52] The ecomodernist argument that there are no fixed boundaries to human consumption aligns neatly with a new wave of extractivism in developing countries with rich but imperiled reserves of biodiversity.[53] One of the defining characteristics of this New Extractivism is the return of the megadam. Hydroelectric power is very much in vogue today, from the Mekong River, where China, Burma, Laos, Thailand, Cambodia, and Vietnam are competing to construct at least 40 big dams, to India's plans to build 160 dams, big and small, on the Brahmaputra River and its tributaries in the northeastern province of Arunachal Pradesh.[54] Support for big dams has grown despite decades of campaigning by grassroots organizations led by people displaced by dams in places such as India's Narmada Valley.[55] These movements argue that big dams have displaced millions of people, generated power that largely benefits small numbers of elites in Global South nations, and had a dramatically negative impact on freshwater ecosystems and biodiversity, assertions supported by the World Bank–convened World Commission on Dams.[56] Underlying these critiques is a challenge to the ideologies of development promoted for the last fifty years by powerful global coalitions of politicians, landed elites, industrialists, international financial institutions like the World Bank, and engineers and scientists.[57] The vision of development advanced by this loose but potent coalition saw development, according to Sanjeev Khagram, as a "large-scale, top-down, technocratic pursuit of economic growth through the intensive exploitation of natural resources."[58] Yet for people's movements around the world, dams are an example of destructive maldevelopment, projects conducted with no regard for social justice and ecological sustainability. Notwithstanding the critiques and campaigns of these social movements, which managed to stall construction of big dams such as those in India's Narmada Valley, a new wave of megadams is slated for development around the globe, with financing justified by the argument that such projects provide access to clean energy.[59]

Nowhere is the folly of the New Extractivism more clearly evident than in the Amazon basin, whose rainforest harbors the world's greatest trove of biodiversity, home to 10 percent of the species on Earth. If humanity were to focus its efforts to fight extinction on one geographical spot, surely the Amazon would be the one. According to the World Wildlife Fund's (WWF) 2014 *State of the Amazon* report, the rainforest and its river basin are home to 40,000 species of plants, 427 different mammals, and 1,300 bird species.[60] Every two and a half square kilometers of rainforest contains at least fifty thousand species of insects. The Amazon's rich biome also stores an estimated 10 percent of global carbon reserves, making the fate of the Amazon a question of vital import for the equilibrium of the entire planet's ecosystems. Contrary to long-standing images of the region of empty wilderness, the Amazon is also densely populated by people—it has one of the highest rates of urbanization in the world—including 385 indigenous groups, many of whom live in protected territories that play a vital role in maintaining the region's environmental sustainability. Yet notwithstanding the region's importance for planetary sustainability, the Amazon's protected natural areas and indigenous territories—the planet's most important harbors of biodiversity—are under threat by development projects, with a wave of downgrading, downsizing, or degazettement carving away protected areas and escalating political and physical threats imperiling indigenous territories and rights. Seventeen percent of the region has been destroyed to date, and, at current deforestation rates of two million hectares per year, a quarter of the rainforest will be destroyed by 2020.[61]

More than sixty big dams and other infrastructural projects are currently slated for the Brazilian Amazon, and neighboring countries are planning scores of dams of their own under a continent-spanning development plan called the Initiative for the Integration of Regional Infrastructure in South America (IIRSA).[62] A dramatically ambitious scheme, IIRSA proposes to weave together South America's major river systems to create an extensive inland canal system linking the Caribbean Ocean to the South Atlantic via the Orinoco, Amazon, Madeira, Paraguay, and Paraná rivers. Once completed, the projected massive infrastructure of big dams and industrial waterways would provide the power and transport required to move vast quantities of resources out of the Amazon, including timber, grain, minerals, and, perhaps most importantly, soybeans, whose cost of transport to markets in East Asia and Europe, where they are used to feed chickens, pigs, and other livestock, would be dramatically decreased. Not surprisingly, IIRSA has received strong political support from mining, logging, and agribusiness interests in countries like Brazil, Peru, Bolivia, and Colombia, but it has also received financial backing from development banks like Brazil's National Bank for Economic and Social Development, from regional entities like the Andean Development Corporation and the Inter-American Development Bank, and from international financial institutions like the UN Development Bank. No attempts have been made to assess the environmental impact of IIRSA, nor have communities that lie in harm's way from the slated projects been informed or consulted about the plans.

The havoc likely to be wrought on the people, flora, and fauna of the Amazon by such development schemes is clearly exemplified by the Belo Monte dam, whose German-built turbines began operating near the end of 2015. Located on the Xingú River in Brazil's Amazon state of Pará, Belo Monte is expected to generate 11,233 megawatts of power, making it the third largest dam in the world in terms of power-generating capacity, after China's Three Gorges Dam and Brazil's own Itaipu Dam on the Paraná River. Once operational, Belo Monte will divert more than 80 percent of the flow of the Xingú River, affecting fish, forests, and navigation along a one hundred kilometer stretch of the river. Millions of fish are likely to die, and several rare species of freshwater fish are expected to go extinct, with dramatic impacts on the indigenous people who live along the river and depend on fishing for their sustenance.

Independent reports have stressed the dubiousness of the Belo Monte project on a number of levels. A panel of engineers estimated, for instance, that during the Amazon's dry season, which lasts a quarter of the year, the dam would generate only a small fraction of its power capacity.[63] In addition, much of the power that the dam will generate is expected to go to fuel the expansion of aluminum smelters and other mineral processing plants in the Amazon rather than to generating accessible and affordable electricity for Brazil's cities, the original legitimation for the project under Brazilian president Lula's Growth Acceleration Program. Even more scandalously, construction on Belo Monte, a project originally initiated during Brazil's brutal U.S.-backed dictatorship, was reinitiated despite court orders paralyzing construction in the 1990s. This was made possible by the Brazilian government's invocation of a legal instrument created during the military dictatorship known as a *suspensão de segurança* or "security suspension." This instrument is predicated on the idea that timetables for dam construction, and the energy they will eventually generate, are more important than the rights of affected populations.

Abrogation of the rights of indigenous people in the Amazon was one of the defining features of Brazil's military dictatorships, which lasted from 1964 to 1985. Military rule was defined by what Paulo Tavares calls "modernizing capitalism," policies which combined authoritarian control in planning and legislation with radical economic liberalism.[64] The high rates of GDP growth achieved under the junta during the 1970s were dependent on a combination of international financial loans, corporate investment, and large-scale exploitation of natural resources for export—a combination that the Belo Monte dam project demonstrates has been revived by the putatively social democratic Workers' Party.[65] This rush to plunder the Amazon was legitimized by the military junta's representation of the region as a "continental void," an attitude that was shared by Brazilian economic and social elites and encouraged by international lending institutions and think tanks such as the World Bank and the Hudson Institute. The exploitation of the Amazon's rich resources was supposed to propel Brazil out of underdevelopment and into the ranks of industrialized nations. In reality, however, the exploitation of the Amazon largely benefited the junta and the mining magnates and livestock barons whose fulsome support it enjoyed. Promises of land for the poor turned out to be largely chimerical, with settlement in the Amazon mainly taking place in the slum-filled peripheries of fast-growing cities like Manaus. The people who suffered the most, however, were Brazil's many indigenous groups.

Under Operation Amazonia, for example, a scheme begun shortly after the coup that brought the military junta to power, the whole region was subjected to a sweeping campaign of resource extraction and intensive agribusiness. Pacification of the indigenous peoples who stood in the way of this scheme was heavily militarized. The junta's repressive policies were given cover by U.S.-based anthropologists such as Napoleon Chagnon, whose best-selling *The Fierce People* (1968) depicted the Yanomami communities as innately violent, thereby legitimating the "civilizing" discourses deployed by the Brazilian military.[66] To counter such discourses, Brazilian anthropologists published clandestine documents such as *The Politics of Genocide against the Indians of Brazil* (1974). Although the term *ecocide* had only recently been coined by scientists opposing the United States's campaign of chemical defoliation in Vietnam and Cambodia, it might equally well have applied to what was unfolding in the Amazon, because Operation Amazonia involved a systematic assault on both the rainforest biome and its human inhabitants. In reaction to a mounting chorus of domestic and international criticism of genocidal policies, the military established FUNAI, the National Indian Foundation, and, in 1973, passed the so-called Indian Statute, which claimed to protect the rights of Amazonian indigenous people. This same law, however, also ruled that "interventions" into indigenous lands were legitimate to "realize public works" or "to exploit the wealth of the subsoil" in the "interest of national security and development."[67] It is this law from the era of dictatorship and genocide that the Workers' Party government invoked to justify continued construction of the Belo Monte dam. True to the history of the bellicose rhetorical nationalism that characterized the junta, Brazil's energy minister defended the Belo Monte dam by calling critics of Belo Monte "demoniac forces that are trying to pull Brazil down."[68]

Faced with the revival of Belo Monte, indigenous groups mounted a defiant campaign that has united tribes who have fought against the specter of "development" for several generations.[69] With only weeks to go before the Belo Monte reservoir begins to flood and the turbines begin to churn, Brazilian federal prosecutor Thais Santi announced a legal action against the dam's builder, Norte Energía, based on the argument that its efforts to squelch indigenous resistance to the dam amount to ethnocide.[70] Unfortunately, Brazil has no existing legislation on ethnocide. Nor does it have any juridical statutes concerning ecocide, although recognition of the rights of nature has been established in Ecuador by a series of landmark legal battles waged by indigenous lawyers in the face of rampant exploitation of the rainforest by transnational corporations such as Chevron/Texaco.[71]

The case against Belo Monte will no doubt be strengthened by emerging revelations concerning corrupt bidding schemes that have enmeshed much of Brazil's economic elite and political class. While the corruption scandal popularly known as Operação Lava Jato, or Operation Car Wash, initially involved billions of dollars' worth of bribes given to government officials and executives in the parastatal oil company Petrobras by members of a cartel of construction companies, jailed executives from this cartel recently promised to expose a similar scheme of massive fraud linked to hydroelectric projects in the Amazon such as Belo Monte.[72] According to initial reports, the cartel of construction companies, influential politicians, and high-level government bureaucrats operated a scheme of bid rigging, bribery, and kickbacks in the Belo Monte and Jirau dam projects. Brazil's Central Accounting Office has also initiated an audit into the billions of dollars' worth of subsidized loans provided by the country's National Development Bank (BNDES) to Norte Energía, loans which covered 80 percent of the construction costs.

Both local and international groups are calling on companies that have been involved in Brazilian hydropower—including well-known corporations such as General Electric, Siemens, Munich RE, EDF Energy, and ENGIE—to reconsider their support for new dams in the delicate Amazonian ecosystem and to instead support truly renewable energy projects. At stake is the habitat for thousands of aquatic and terrestrial species as well as the livelihoods of indigenous people such as the Munduruku. With a corruption-riddled new government in power with strong ties to agribusiness and development interests, the fate of both human and non-human forest dwellers is tenuous. Much will depend on the ways in which indigenous groups and environmental organizations can galvanize support against the corrupt forms of development that revived Belo Monte.

CONCLUSION

"Islands," in the words of Elizabeth Kolbert, "tend to be species-poor, or, to use the term of art, depauperate."[73] If real islands in the world's oceans are relatively low in biodiversity, the same holds true of isolated patches of wilderness such as those preserved by the so-called Biological Dynamics of Forest Fragments Project, which Kolbert profiles in her book on extinction. In a vast experiment in the Amazon, tracts of rainforest were left standing while everything around them was clear cut. Biologists then monitored the biodiversity in the preserves. What they found was not a new equilibrium with fewer species but rather a steady degradation of biodiversity over time.[74] Isolated patches of wilderness, in other words, cannot abide indefinitely in the midst of a gale of capitalist destruction. Capitalism depauperates the planet.

Moreover, there is what Kolbert calls a "dark synergy" between habitat fragmentation and climate change: as the climate warms, species need to migrate to new environments to survive, but "development" is increasingly creating barriers—roads, clear-cuts, cities—that impede mobility and isolate species in fragmentary environments.[75] The implication is that conservation, which is predicated on preserving isolated patches of "wilderness," will ultimately fail unless the broader system of capitalist development that fragments habitats is not challenged. At the moment, precisely the opposite process is playing out. Indeed, the very organizations charged with conservation have internalized neoliberal capitalist models. The UN Convention on Biological Diversity, for instance, launched in 2008 a model for marketing "environmental services" through the UN Framework Convention on Climate Change's (UNFCCC) Business and Biodiversity Initiative, which includes mechanisms for offsets and for the creation of "natural capital."[76] Within such schemes, the environmental commons of the Global South, the planet's tropical forests and oceans, and the myriad creatures who inhabit them become a source of natural capital that can be quantified and traded on global markets. Biodiversity is thereby transmuted into a source of offset credits that allow polluting corporations and governments to continue their ecological mayhem. Some of the world's most prominent conservation-based environmental non-governmental organizations have signed on to this disaster biocapitalism, including Conservation International, the Worldwide Fund for Nature, the Nature Conservancy, and the Environmental Defense Fund.[77]

If mainstream environmentalism has been coopted by such neoliberal policies, what would an anticapitalist movement against extinction look like? It would begin from the understanding that the extinction crisis is at once an environmental issue *and* a social justice issue, one that is linked to long histories of capitalist domination over specific people, animals, and plants. The extinction crisis needs to be seen as a key element in contemporary struggles against enclosures. The extinction crisis, in other words, ought to be a key issue in the fight for climate justice. If techno-fixes such as de-extinction facilitate new rounds of biocapitalist accumulation, an anticapitalist movement against extinction must be framed in terms of a refusal to turn land, people, flora, and fauna into commodities. An anticapitalist movement against extinction must reject capitalist biopiracy and imperialist enclosure of the global commons, particularly when they clothe themselves in arguments about preserving biodiversity. Forums for enclosure like the UNFCCC's Business and Biodiversity Initiative must be recognized for what they are and shut down. Most of all, an anticapitalist movement must challenge the privatization of the genome as a form of intellectual property, to be turned into an organic factory for the benefit of global elites. Synthetic biology must be regulated.[78] The genomic information of plants, animals, and human beings is the commonwealth of the planet, and all efforts to make use of this environmental commons must be framed around principles of equality, solidarity, and environmental and climate justice.

NOTES

1. Tom Roston, "Illuminating the Plight of Endangered Species, at the Empire State Building," *New York Times,* July 29 2015, http://www.nytimes .com/2015/07/31/movies/illuminating-the-plight-of-endangered-species-at-the -empire-state-building.html?_r=1.

2. Elizabeth Kolbert, *The Sixth Extinction* (New York: Henry Holt, 2014), 266.

3. Ibid.

4. Mark Dowie, *Conservation Refugees: The Hundred-Year Conflict between Global Conservation and Native Peoples* (Cambridge, Mass.: MIT Press, 2009).

5. Rosaleen Duffy, *Nature Crime: How We're Getting Conservation Wrong* (New Haven, Conn.: Yale University Press, 2010).

6. Mark Dowie, "Conservation Refugees," *Orion Magazine,* https://orion magazine.org/article/conservation-refugees/.

7. Ibid.

8. Ramachandra Guha and Juan Martinez-Alier, *Varieties of Environmentalism: Essays North and South* (London: Earthscan, 1997), 12.

9. James O'Connor, *Natural Causes: Essays in Ecological Marxism* (New York: Guilford Press, 1997), 166.

10. Joseph Schumpeter, *Capitalism, Socialism, and Democracy* (New York: Harper Perennial, 2008).

11. See David Harvey, *Seventeen Contradictions and the End of Capitalism* (New York: Oxford University Press, 2015).

12. On the ethical and political implications of emerging understandings of the world as composed of multiple interacting systems, see William E. Connolly, *A World of Becoming* (Durham, N.C.: Duke University Press, 2011).

13. Elizabeth Kolbert, "Recall of the Wild: The Quest to Engineer a World Before Humans," *The New Yorker,* December 24, 2012, http://www.newyorker .com/magazine/2012/12/24/recall-of-the-wild.

14. George Church and Ed Regis, *Regenesis: How Synthetic Biology Will Reinvent Nature and Ourselves* (New York: Basic Books, 2012), 10.

15. On the transformation of life into code, see Eugene Thacker, *The Global Genome: Biotechnology, Politics, and Culture* (Cambridge, Mass.: MIT Press, 2005).

16. Church and Regis, *Regenesis,* 10.

17. Nathaniel Rich, "The Mammoth Cometh," *New York Times Magazine,* February 27, 2017, https://www.nytimes.com/2014/03/02/magazine/the-mammoth-cometh.html?_r=0.

18. Church and Regis, *Regenesis,* 12.

19. Ibid.

20. Josh Donlan, "De-extinction in a Crisis Discipline," *Frontiers of Biogeography* 6, no. 1 (2014): 25–28.

21. Editorial, "Why Efforts to Bring Extinct Species Back from the Dead Miss the Point," *Scientific American* 308, no. 6 (May 14, 2013), http://www.scientificamerican.com/article/why-efforts-bring-extinct-species-back-from-dead-miss-point/.

22. Norman Carlin, Ilan Wurman, and Tamara Zakim, "How to Permit Your Mammoth: Some Legal Implications of 'De-Extinction,'" *Stanford Environmental Law Journal* 33, no. 1 (2014): 3–57, https://journals.law.stanford.edu/stanford-environmental-law-journal-selj/print/volume-33/number-1/how-permit-your-mammoth-some-legal-implications-de-extinction.

23. Melinda Cooper, *Life as Surplus: Biotechnology and Capitalism in the Neoliberal Era* (Seattle: University of Washington Press, 2008), 27.

24. Ibid., 23.

25. Church and Regis, *Regenesis,* 4.

26. Melinda Cooper makes this extremely productive link between the monetarist revolution and the growth of research funding for the life sciences. See Cooper, *Life as Surplus,* 29–31.

27. David Harvey, *The New Imperialism* (New York: Oxford University Press, 2003), 67.

28. Ilya Prigogine and Isabelle Stengers, *Order out of Chaos: Man's New Dialogue with Nature* (New York: Bantam, 1984).

29. Cooper, *Life as Surplus,* 38.

30. Naomi Klein, *The Shock Doctrine: The Rise of Disaster Capitalism* (New York: Picador, 2008).

31. Stefano Liberti, *Land Grabbing: Journeys in the New Colonialism* (New York: Verso, 2014).

32. An Ecomodernist Manifesto, http://www.ecomodernism.org/.

33. Ian Angus, "Hijacking the Anthropocene," *Monthly Review,* May 19, 2015, https://mronline.org/2015/05/19/angus190515-html/.

34. *An Ecomodernist Manifesto,* April 2015, 6. Further page references are provided in the text.

35. Stewart Brand, *Whole Earth Discipline: Why Dense Cities, Nuclear Power, Transgenic Crops, Restored Wildlands, and Geoengineering Are Necessary* (New York: Penguin, 2010).

36. Harvey, *Seventeen Contradictions,* 222–45.

37. Tim Jackson, *Prosperity without Growth: The Transition to a Sustainable Economy* (London: Sustainable Development Commission, 2009), 49, http://www.sd-commission.org.uk/publications.php?id=914.

38. Ibid., 50.

39. Ibid., 54.

40. Richard Barbrook and Andy Cameron, "The Californian Ideology," *Science as Culture* 6, no. 1 (1996): 44–72, archived at Imaginary Futures, http://www.imaginaryfutures.net/2007/04/17/the-californian-ideology-2/.

41. Newt Gingrich, *A Contract with the Earth* (Baltimore: Johns Hopkins University Press, 2007); Paul Hawken and Amory Lovins, *Natural Capitalism: Creating the Next Industrial Revolution* (London: Earthscan, 1999).

42. Stewart Brand, "Rethinking Extinction," *Aeon,* April 21, 2015, https://aeon.co/essays/we-are-not-edging-up-to-a-mass-extinction.

43. Ibid.

44. Editorial, "Protect and Serve," *Nature* 516, no. 7530 (2014), http://www.nature.com/news/protect-and-serve-1.16514.

45. Ibid.

46. Ibid.

47. Ashley Dawson, "Save the Mosquito," *Jacobin,* April 22, 2016, https://www.jacobinmag.com/2016/04/zika-mosquitos-crispr-genetic-engineering-puerto-rico/.

48. Ramachandra Guha and Juan Martinez-Alier, *Varieties of Environmentalism: Essays North and South* (London: Earthscan, 1997), 12.

49. *An Ecomodernist Manifesto,* 17.

50. On the destructive history of development, see Gilbert Rist, *The History of Development: From Western Origins to Global Faith* (New York: Zed, 1996), and Karin Kapadia, ed., *The Violence of Development: The Politics of Identity, Gender, and Social Inequalities in India* (New York: Zed, 2002).

51. Michael Shellenberger and Ted Nordhaus, "Statement on 'Climate Pragmatism' from BTI Founders Michael Shellenberger and Ted Nordhaus," *The Breakthrough* (blog), July 27, 2011, http://thebreakthrough.org/archive/statement_on_climate_pragmatis.

52. *An Ecomodernist Manifesto,* 20.

53. On this new wave of resource exploitation, see Henry Veltmeyer and James Petras, *The New Extractivism: A Post-Neoliberal Development Model or Imperialism of the Twenty-First Century?* (New York: Zed, 2014).

54. On the Mekong dams, see Michelle Nijhuis, "Harnessing the Mekong, or Killing It?," *National Geographic,* May 2015, http://ngm.nationalgeographic.com/2015/05/mekong-dams/nijhuis-text. On dams in India's northeastern provinces, see Kieran Cooke, "The Dams of India: Boon or Bane?," *The Guardian,* March 17, 2014, http://www.theguardian.com/environment/2014/mar/17/india-dams-rivers-himalaya-wildlife.

55. On opposition to big dams in India, see Sanjeev Khagram, *Dams and Development: Transnational Struggles for Water and Power* (Ithaca, N.Y.: Cornell University Press, 2004).

56. World Commission on Dams, *Dams and Development: A New Framework for Decision-Making* (London: Earthscan, 2000), https://www.internationalrivers.org/resources/dams-and-development-a-new-framework-for-decision-making-3939.

57. Khagram, *Dams and Development,* 4.

58. Ibid.

59. For social movement critique of dam industry greenwashing, see International Rivers, "Activists Protest Corporate Greenwashing of Dams at World Water Forum," *International Rivers* (blog), March 14, 2012, http://www.interna tionalrivers.org/resources/activists-protest-greenwashing-of-dams-at-world -water-forum-3683.

60. C. C. Maretti, S. J. C. Riveros, R. Hofstede, D. Oliveira, S. Charity, T. Granizo, C. Alvarez, P. Valdujo, and C. Thompson, *State of the Amazon: Ecological Representation in Protected Areas and Indigenous Territories* (Brasília: WWF Living Amazon [Global] Initiative, 2014), 7, http://wwf.panda.org/what_we_do /where_we_work/amazon/?232871/State-of-the-Amazon-Ecological-Represen tation-in-Protected-Areas-and-Indigenous-Territories.

61. Ibid., 8.

62. For information on IIRSA, see "Amazônia Viva," *International Rivers* (blog), undated, http://www.internationalrivers.org/campaigns/amazônia-viva.

63. "Independent Review Highlights the True Costs of Belo Monte Dam," *International Rivers* (blog), October 12, 2009, http://www.internationalrivers .org/resources/independent-review-highlights-the-true-costs-of-belo-monte -dam-3783.

64. Paulo Tavares, "The Geological Imperative," in *Architecture in the Anthropocene: Encounters among Design, Deep Time, Science, and Philosophy,* ed. Etienne Turpin (London: Open Humanities Press, 2013), 212.

65. Ibid.

66. Ibid., 213.

67. Ibid., 226.

68. "Independent Review." On the use of nationalist rhetoric by Brazilian elites against environmentalist and indigenous critics, see Andrea Zhouri, "'Adverse Forces' in the Brazilian Amazon: Developmentalism versus Environmentalism and Indigenous Rights," *The Journal of Environment and Development* 19, no. 3 (2010): 252–73.

69. Jonathan Watts, "Amazonian Tribes Unite to Demand Brazil Stop Hydroelectric Dams," *The Guardian,* April 30, 2015, http://www.theguardian.com /world/2015/apr/30/amazonian-tribes-demand-brazil-stop-hydroelectric-dams, and Jonathan Watts, "The Tribes Living in the Shadow of a Megadam," *The Guardian,* December 16, 2014, http://www.theguardian.com/environment/2014 /dec/16/belo-monte-brazil-tribes-living-in-shadow-megadam.

70. Mario Osava, "Indigenous Peoples in Brazil's Amazon—Crushed by the Belo Monte Dam?," *Amazon Watch* (blog), July 16, 2015, http://amazonwatch .org/news/2015/0716-indigenous-people-in-brazils-amazon-crushed-by-the-belo -monte-dam.

71. See Ursula Biemann and Paulo Tavares, *Forest Law* (2014, video installation), Bridging Art and Knowledge gallery, Utrecht, Netherlands, 2015, https://

www.bakonline.org/en/Research/Itineraries/Future-Vocabularies/Themes/Human-Inhuman-Posthuman/Exhibitions/Forest-Law.

72. "Massive Corruption Scandal Implicates Brazil's Dam Builders," *International Rivers* (blog), March 4, 2015, http://www.internationalrivers.org/resources/8595.

73. Kolbert, *Sixth Extinction,* 179.

74. Ibid.

75. Ibid., 189.

76. Global Justice Ecology Project, *The Green Shock Doctrine* (Buffalo, N.Y.: Global Justice Ecology Project, n.d.), 5, http://globaljusticeecology.org/green-shock-doctrine/.

77. Ibid., 8.

78. A promising initial step in this direction was taken during the UN Convention on Biological Diversity meeting in 2014, although it met fierce opposition from nations, such as the United States and the United Kingdom, with strong synthetic biology industries. See SynBio Watch, "Regulate Synthetic Biology Now: 194 Countries," *SynBio Watch* (blog), October 20, 2014, http://www.synbiowatch.org/2014/10/regulate-synthetic-biology-now-194-countries/.

9

Surviving the Sixth Extinction: American Indian Strategies for Life in the New World

Daryl Baldwin, Margaret Noodin,
and Bernard C. Perley

Is there life after extinction? The authors of this chapter argue yes. At first glance, the assertion that "there is life after extinction" seems trivial because some species have survived earlier mass extinctions. Humans are the beneficiaries of earlier extinctions. Yet, today, new anxieties regarding the survivability of the human species proliferate in this current era that many proclaim as the "sixth extinction." The current fascination regarding whether we are in the midst of the sixth extinction is popularized in trade publications, symposia, and popular media.[1] Richard Leakey asserts, "I will state boldly right now that I believe we face a crisis—one of our own making—and if we fail to negotiate it with vision, we will lay a curse of unimaginable magnitude on future generations."[2] This extinction crisis has fomented catchphrases such as "politics of risk" and "an age of risk" and popular imaginings such as "odds against tomorrow."[3] This contemporary zeitgeist of uncertainty has spread from environmental and biological sciences to social sciences. This chapter addresses parallel rhetoric and interventions from linguistics and related language sciences.

The crisis rhetoric that language scientists use to alert the general public to the immediate global linguistic crisis includes "vanishing voices," "last speakers," and "dying words."[4] The arguments for saving endangered languages are similar to those used by the sciences for saving endangered ecosystems and species. Language scholars assert that when a language becomes extinct, we all lose, because that language represents part of our human heritage, and language loss is loss of biocultural diversity as well as loss of language ecologies.[5] Echoing Leakey's "crisis of our own making," language scholars also argue that the global linguistic crisis is linked to the colonial processes that altered the biocultural environments where heritage languages flourished before colonial settlement. Ironically, contemporary language scholars, while sounding the alarm, may be exacerbating the problem for many endangered languages. The impulse to document the last words of the last speakers is appropriate for languages with a handful of elderly speakers. For many endangered language communities, the salvage sciences represent a salvage-through-documentation solution that is deaf to the needs of heritage language communities. The authors of this chapter offer perspectives that may enhance the survival of those languages. The authors also offer insights into what surviving mass extinction would look like.

MASS EXTINCTION COMES TO AMERICA

Leakey warns that we are laying "a curse of unimaginable magnitude on future generations." This statement is both prophetic and descriptive. The sixth extinction soothsayers are concerned with the fate of humans in the aftermath of mass extinction. What are the conditions that will allow humans to survive such a cataclysm? What are the criteria for proclaiming species/entities/environments to be extinct? Is "extinction" a self-fulfilling prophecy? In cases of extinction, what technological interventions present opportunities for reanimation, revitalization, or de-extinction? The authors argue that the conditions required for surviving mass extinction are in part determined by the kinds of questions we ask, the metaphors we use, and the actions we take.

   The authors argue that contemporary native North America is an example of surviving the first convulsion of the sixth extinction. Our current global extinction, humanity's "first extinction," has been a slowly evolving global disaster, and recent observations and analyses indicate dramatic increases in the extinction of species worldwide. Human

Figure 9.1. Bernard Perley, "Unpacking Colonial Baggage in Native North America," panel 1 of 3, from the *Having Reservations* series. The complete cartoon appears in Appendix A. Translations of the cartoon into Myaamia, Maliseet, and Anishinaabemowin appear in Appendix B. Courtesy of the artist.

activities are attributed to the causes, direct and indirect, of the current species extinction crisis. The critical concern for our sixth extinction is the disruption in sustaining human longevity. Our species survival may be at stake in the unfolding tragedy of our sixth extinction. Yet, there may be perspectives that offer some optimism. Surviving mass extinction is dependent on alternative perspectives that promote "awakening," "emergent vitalities," and new configurations of "sovereignty" in the face of "language extinction." The authors of this chapter offer experience in mass extinction survival. Each author has had to deal with human activities that contribute to the extinction, endangerment, and misrepresentation of their languages, cultures, and communities. The "sixth extinction" has a start date for native North America: October 12, 1492.

The Beginning of the End?

On October 12, 1492, Christopher Columbus initiated a series of events that would lead to the extinction of most of the American Indian languages, cultures, and communities of North America. Columbus's first act was to take possession of the island Guanahaní for the Spanish Crown by "making the requisite declarations, as is more fully recorded in the statutory instruments which were set down in writing."[6] When these declarations were completed, "many islanders gathered round."[7] Columbus's *Journal of the First Voyage* provides a long paragraph describing his first encounter with the native peoples of the island Guanahaní. At one point, Columbus makes an observation that would have devastating consequences for the American Indian peoples in the ensuing five centuries:

> They ought to make good slaves for they are of quick intelligence since I notice that they are quick to repeat what is said to them and I believe that they could very easily become Christians, for it seemed to me that they had no religion of their own. God willing, when I come to leave I will bring six of them to Your Highnesses so that they may learn to speak.[8]

It is a curious ethnographic account that reveals more than just a descriptive observation. It also reveals an ideology of conquest, empire, and self-serving superiority. Five centuries later, the surviving American Indian peoples are grappling with the legacies of colonialism that continue to undermine the vitality of their languages, cultures, and religious practices. Among those oppressive colonial legacies are concepts of extinction and endangerment. This chapter frames these concepts as "invasive concepts" that perpetuate colonial domination and oppression of indigenous concepts and practices. Language experts have used discourses and rhetoric to reify their gloss of extinction as authoritative pronouncements eliciting particular forms of intervention. American Indian language advocates have been pushing back to shift the focus away from extinction and toward vitalities emerging from communities of continuity. It is not enough to challenge colonial concepts: there has to be a solution.

The End of the Beginning

This chapter presents three case studies of language vitality that challenge Western notions of extinction and endangerment. The first case is the most dramatic in terms of challenging the concept of *extinction*.

Daryl Baldwin of the Miami Tribe of Oklahoma describes the vitalities that emerge from changing the expert rhetoric away from "extinction" to "sleeping."[9] The second case study presents Bernard Perley's work toward emergent vitalities in the Maliseet community of Tobique First Nation. Language experts have diagnosed the Maliseet language as severely endangered. Endangerment implies a trajectory based on statistical data regarding language as the object of inquiry. Perley argues that the emphasis on the linguistic code is misplaced and that more attention must be paid to the community and the emerging social relationships that are mediated by the Maliseet language. The third case study is an exercise in repatriating linguistic sovereignty across national borders as well as U.S. state and Canadian provincial borders. Margaret Noodin describes how her work to promote Anishinaabemowin language instruction in the Great Lakes watershed is repatriating traditional Anishinaabe sovereign lands. In all three case studies, the intimate links between language, culture, religion, and landscape are highlighted. These are integrated cultural resources that defy colonial notions of knowledge, agency, and power.

Language experts continue to place emphasis on their linguistic sciences as unquestioned modes of analysis and practical intervention. Their rhetoric of loss and extinction is drawn from faulty metaphors that misguide these experts into overlooking other perspectives that may be instrumental in accomplishing their goals of language maintenance and revitalization. The importance of shifting from language death to language life will provide American Indian communities with affirmation and optimism for possible "linguistic" futures. The experts have also constrained the kinds of agency that "endangered language" communities can exercise. Endangerment and extinction perpetuate the victimization of American Indian communities and the domination of colonial "knowledge" over American Indian "experience." The endangerment-to-extinction trajectory may serve to "sound the alarm" for language experts and potential funding sources, but it also diminishes the agency of the language communities. Shifting the rhetoric to "vitality" allows communities to see the promise and potential in their respective languages and thereby provides members with agency and mutually coordinated futures. Finally, the emphasis on the experts and their rhetoric reinstitutes the colonial institutional power over American Indian communities. The funding agencies, the learned institutions, and the experts themselves all enjoy the prestige and privilege of colonial regimes of power. Even when the expert discourses acknowledge their power positions, they can do so because they can afford that luxury. Their position as experts is protected by colonial states. However, there is a growing movement toward repatriating American Indian sovereignty, decolonizing knowledge, re-creating cultural confederacies, and reenergizing native networks. As the three authors point out in their respective case studies, sovereignty can be exercised and experienced in many different ways and in many different contexts. This chapter provides only three such sovereign experiences.

EXTINCTION BY INVASION

Invasive Species and Mass Extinction

It is not a coincidence that the areas where there are the greatest number of indigenous languages "dead or dying" are North America, at a rate of 61 percent, and, jointly, Australia and New Zealand, at a rate of 82 percent.[10] Gary Simmons and M. Paul Lewis explain that the severe rates of extinction are due to colonization by settlement. Drawing from Salikoko Mufwene's 2002 article on colonization and globalization, they conclude that "the places where language loss has been the most profound—Australia, Canada, and the United States—are all places where virtually all of the land was settled by the colonizers, thus displacing the indigenous inhabitants."[11] The arrival of colonizers was not the only contributing factor to the extinction of indigenous languages. The introduction of new species of animals, plants, pathogens, microorganisms, and so forth had precipitated the eradication of stable indigenous biogenic processes.[12] The extinction of integrated species across human, biological, and geological domains is mass extinction that "has 'reset' the global biological system, in the sense that important groups of organisms have disappeared, making way for the expansion of others."[13] Though speaking from the perspective of paleontology, Steven Stanley argues that "mass extinctions resulted from certain other, more mundane causes."[14] From the perspective of colonial settlers in North America, Australia, and New Zealand, it was a mundane act to import foreign plants and animals from their "old worlds" into their "new worlds." From the perspective of the native populations, this invasion of species was not mundane. The settlers changed the landscape with intensive farming, grazing, and settlement patterns that reconfigured traditional indigenous homelands into alien landscapes. Coupled with the devastating effects of old world pathogens on native communities and the disruptions to generations, continuity in transmission of traditional knowledge and forced displacement contributed to widespread collateral extinctions.[15] Those collateral extinctions include American Indian languages, "education, religion, knowledge, everyday social interactions, and identity."[16] Not only are indigenous languages at risk of extinction, so are the social relations that are mediated by those languages. Those social relations were disrupted by settler societies and abruptly replaced by invasive concepts that are now being examined for the impact they have had on American Indian communities. Among those concepts are "extinction" and "endangerment."

Efficacy of Invasive Concepts

Language experts estimate that anywhere from 50 to 90 percent of the world's languages will become extinct by the end of the twenty-first century.[17] Some scholars use the metaphor of language as biological organism while also pointing out the limitations of the metaphor.[18] The perceived benefit of the metaphor is the way it highlights the crisis aspect of their discipline and thereby enhances their appeal for support to fund language salvage projects. The metaphor also promotes particular interventions to "save" dying or endangered languages. Those interventions include documentation of all aspects of language use within the endangered language community. The end products are usually material artifacts such as texts, audio and video files, digital archives, interactive websites, and other modes of material representations of language. For many severely endangered languages, these modes of documentary linguistics are necessary practices to salvage what is left of a dying language before the last speaker breathes his or her last breath. However, these documentation strategies may be inappropriate for other language states.

Combining the concepts of extinction and endangerment with the metaphor of biological organism to describe the state of any human language provides both practical solutions and unintended constraints. The unintended constraints, specifically the focus on documenting and archiving the linguistic code, ironically contribute to the further destabilization of those languages. The insistence that expert and community efforts should be dedicated to documentation leads to reallocating time and resources to mechanical interventions instead of human interventions. In short, the metaphors present not only partial (biased) representations of language and social relations but also partial (incomplete) modes of language preservation.[19] For those languages with a community of language advocates who continue to use their heritage languages in traditional and innovative ways, documentation is not enough. The survival of their languages rests on repatriating the terms, practices, and ideologies of American Indian linguistic sovereignty.

At the end of the nineteenth century, after four hundred years of species loss, American Indian population contraction (according to the 1890 U.S. census, only a little over 250,000 American Indians were counted in the survey), language and cultural eradication, extreme landscape modification, and the "end of the frontier" extinction by invasion all moved toward a subtler but equally effective mode of background extinction for native North America.[20] When the last remnants of Indian land were bounded, tamed, and/or appropriated, the only task left was to "kill the Indian and save the man."[21] In Canada, the deputy superintendent of Indian affairs, Duncan Campbell Scott, declared, "I want to get rid of the Indian problem. . . . Our objective is to continue until there is not a single Indian in Canada that has not been absorbed into the body politic."[22] The surviving American Indian populations were subjected to programs of assimilation, relocation, and religious conversion. Columbus's original impulse to deny American Indians their sovereignty continued unabated at the beginning of the twentieth century. However, while these assaults on American Indian ontologies undermined indigenous traditions and continuities, a growing network of scholars and advocates feared that the Indians were about to vanish. The vanishment thesis would lead to the development of salvage efforts by scholars and concerned citizens as they went to Indian country to preserve what was left before it all disappeared.[23] In Milwaukee, Wisconsin, where Anishinaabemowin is taught in elementary classes, on the university campus, and at weekly community gatherings, these efforts defy the predictions once made by native and nonnative Americans. As historian Robert Trennert reminds us, the removal and reservation eras, followed by federal termination policies, allowed settlers to view American Indians as a "vanishing race." In 1851, *The Milwaukee Daily Free Democrat,* edited by abolitionist Sherman Booth, published this warning: "It is melancholy to meditate upon the sure extinction of the Aboriginal inhabitants of America. Still such is indisputably the fact, that the last remnant of this race will, in a few years, become extinct."[24] Even Potawatomi leader and author Simon Pokagon explained that he included words and phrases in Potawatomi throughout his novel, *Ogimawkwe Mitigwaki* (queen of the woods), "in consideration of the fact that the language of the great Algonquian family, which was spoken by hundreds of thousands throughout more than half of North America, is fast passing away."[25] This "vanishment" was intended as an act of revisionism that was "irrevocable, complete, and irreversible."[26] In the Miami case, the vanishment sentiment was echoed in the following: "Thus shrouded in darkness, with the lights of civilization and religion beaming around them, the last fragments of one of the most powerful aboriginal nations of North America are passing away from the earth forever. The arms of the Miamies are now powerless. Their last lingering bands are slowly tottering towards the grave of their nation."[27] Vanishment also doomed the Maliseet into extinction. John Dyneley Prince writes, "I cannot see meaning in the word *Wabanaki,* 'land of the dawn or east,' which points to any period further back than the time of these peoples' first tribal centralization on the present eastern coast of North America. Let then our labor in this work suffice merely to present to the English-speaking public a few interesting and characteristic specimens of the traditions of a rapidly perishing race—a race which fifty years from now will have hardly a single living representative."[28] The nineteenth-century mass extinction crisis for native North America produced new epistemologies of expertise and intervention. Those epistemologies continued to assert colonial power over American Indian experiences in favor of salvage science. Rather than conquest by force, American Indians had to endure conquest by ideology from two directions: assimilation and salvage science. The legacy of early-twentieth-century interventions would be repeated in the early decades of the twenty-first century. Today, despite population recovery for American Indian communities, we are witnessing the continued assault on American Indian ontological self-determination. The new salvage science uses the same methods of documentation and expert judgment. Perhaps it is time to unpack all that colonial baggage.

UNPACKING COLONIAL BAGGAGE

The unfortunate consequences of colonial epistemological invasions are the perpetuation of victimization of American Indian peoples and the reification of colonial power. Today, the language experts are the self-assigned saviors of American Indian languages. Their rhetorical strategy to use "language extinction" as the operative metaphor for identifying a linguistic crisis has brought needed attention and support for language documentation projects as well as institutional support from universities, nongovernmental organizations, and language communities. These are all positive developments with tangible benefits. But, is this call to action for the communities whose languages are diagnosed as endangered or near extinction? Is linguistic science the primary beneficiary for salvage science? In regard to native North America, will American Indians continue to be victimized by settler epistemologies and subtler reifications of colonial power? The authors offer perspectives that could realign epistemic stances between language experts and community language activists that promote mutual respect and establish a common ground that assures American Indian language, cultural, and sovereign vitalities. However, the first step to repatriating American Indian linguistic sovereignty is to recognize the deleterious aspects of Western concepts for indigenous cultures and communities.

Extinction is not only a foreign concept but an invasive one. When language experts use *extinction* to describe the state of American Indian (or any other) language, they borrow the concept from biology. The definition of *extinction* is as follows: "1: the act of making extinct or causing to be extinguished; 2: the condition or fact of being extinct or extinguished; *also*: the process of becoming extinct <*extinction* of a species>; 3: the process of eliminating or reducing a conditioned response by not reinforcing it."[29]

This definition includes process in the description along with condition or state of being. This is a significant aspect of the definition, as it also

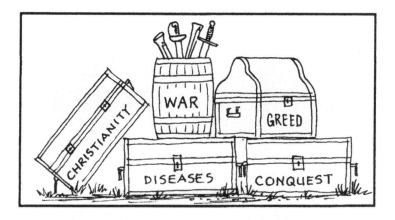

Figure 9.2. Bernard Perley, "Unpacking Colonial Baggage in Native North America," panel 2 of 3, from the *Having Reservations* series. The complete cartoon appears in Appendix A. Translations of the cartoon into Myaamia, Maliseet, and Anishinaabemowin appear in Appendix B. Courtesy of the artist.

describes the unwitting complicity of language experts in promoting language extinction through their use of biological metaphors. Daryl Baldwin points out in the Myammia case that the term *extinction* was interpreted by tribal citizens as "gone forever."[30] If Baldwin had not challenged the expert diagnosis, he would not have been able to initiate the Myaamia awakening efforts. Furthermore, the authors have had conversations regarding the word *extinction* as an invasive concept. Daryl Baldwin states that "David Costa and I have discussed this several times and he was not able to find solid equivalents for the concept of 'extinction' in any of the known Algonquian languages, but in some cases speakers used something that translates to 'absent/gone' to express the idea. So yes, I do think it's a foreign concept in the context of how it gets used in situations like this."[31] Coauthor Margaret Noodin adds, "Our concept in Anishinaabemowin centers on 'jaag'—to exhaust, use up, or completely destroy something or someone." A basic example would be the way it is used by Duncan Pegahmagabow to state that a basic supply has been exhausted:

> *Naajbatwaadin iihow ay'ii ziisbaakwad ngii-jaagsemin kaa geyaabi gegoo ziisbaakwad.*

Run home and fetch some sugar, we have run out and there is no more sugar here.[32]

Or as it was used by Andrew Medler to say,

Eshkam znagad wii-debnaming iw waaboodweng. Wjaaggahaanaawaa niibna wdakiimwaan giw Nishnaabeg.
It is getting harder and harder to obtain firewood. Those Indians are clearing much of their land.[33]

In a more intense example, Sam Osawamick used the term when speaking of a specific battle during the War of 1812:

Mii sa go eta kina gii-jaagnanaawaad wadi biindig eyaan'jin.
So they, alone, killed all those that were inside.[34]

Another term offered by Anishinaabemowin teacher Alphonse Pitawa-nakwat is *nagawanaagawad,* which means that something inanimate has disappeared completely or will never be seen again. *Gaawiin maashi An-ishinaabemowin nagawanaagawasinoon* (Anishinaabemowin has not yet disappeared).[35]

In regard to the Maliseet, Perley consulted with community-recognized Maliseet language speaker and elder Henrietta Black to determine whether Maliseet had terms for the concept of extinction. The nearest approximation was *kikatahasu,* literally meaning "it's all killed off." One other option was *pektahasu,* meaning "wiped out." This term was explained in distinction from "erased"—that term would be *kahsihutasu.*[36]

Though the three languages cited here constitute a small sample size, the discussion does reinforce the need to critically evaluate the discourses of the language experts and identify how those discourses conceal concepts and actions that contribute to the very condition that they are trying to alleviate. The translations from Myaamia, Anishinaabemowin, and Maliseet into English are not perfect glosses for the English word *extinction* because each language would regard using its respective term to describe a state or condition of its heritage language as an anomalous utterance. The semantic properties of the American Indian words denoting *extinction* do not include *language* in their semantic fields. Ironically, language experts' use of *extinct* and *endangerment* is also a case of an anomalous utterance. Only when the expressions are marked as metaphors can the semantic properties make any meaningful contribution to understanding. However, even if we do acknowledge expanding the semantic domains for *extinct* and *endangerment* to include language, we must also realize that the descriptor will always be partial. It is partial in two important ways. First, it is partial because all metaphors are incomplete by their very nature. Second, it is partial because it reflects the perceptual stance of the language experts. That, in turn, influences the kinds of interventions those experts promote to "save" endangered languages. Unfortunately, from this perspective, because extinction is a permanent state of affairs, nothing can be done for extinct languages. Therein lies the critical work of American Indian language advocates such as Daryl Baldwin. He did not accept the invasive concept *extinction*. Baldwin and others exercised Myaamia linguistic sovereignty by changing the metaphor from *extinct* to *sleeping*.[37] That led to the possibility that the Myaamia language can be awakened. Part of that awakening is the concerted effort to anchor the language awakening program in the stories, the landscape, the relationships, that are integral to Myaamia heritage.[38] Baldwin is not alone in his effort to rethink Western concepts and the consequences of those concepts for the survival of American Indian languages. Together, all three authors provide examples of indigenous language advocacy that challenge the invasive concepts that arrived in the Americas as part of colonial baggage.

THREE MODELS OF AMERICAN INDIAN LINGUISTIC SOVEREIGNTY

The relative states of language vitality for three specific Algonquian languages will be the focus of discussion. In each we find similarities and differences and new ways to think about language vitality. The spectrum of language vitality that will be discussed begins with "extinct" languages (e.g., Miami-Illinois [Myaamia]) and then moves to "severely endangered" languages (Maliseet-Passamaquoddy, Anishinaabemowin).[39]

Miami-Illinois (Myaamia)

The Miami-Illinois language is a central Algonquian language spoken historically in the lower Great Lakes region. The historical homelands encompass what are now the states of Illinois and Indiana, western Ohio, and southern portions of Michigan and Wisconsin. This landscape was shared with many other tribal villages and languages, most of whom had a kinship relationship with the Miami-Illinois and equally claim this region as historic territory.

Beginning in 1846, the Miami Tribe (as a self-governing entity) was forcibly relocated west of the Mississippi to the unorganized territory and placed on a three hundred thousand acre reservation. Several exempt families remained behind, thereby fragmenting the population. By 1870, the Kansas reservation was allotted, and again the Miami Tribe was moved to Indian Territory (now Oklahoma), leaving behind many exempted families in Kansas. This constant relocation, community fragmentation, and continued social pressures to "conform" to American society severely weakened their ability to maintain important aspects of their identity as Myaamia people, including their language. By the mid-twentieth century, the last speakers were passing, leaving behind a great deal of cultural shame and uncertainty about their future.

A notion like *extinction* is best communicated in the Myaamia language from a biological context. For instance, it would be considered very acceptable discourse to say something like *ceekineeciki eelikwaki,* "all the ants are dead." The verb in this statement is *ceekinee,* "be all dead," but of course in this context it doesn't mean extinct, because we know that not all ants are dead. It would also be very acceptable to say *ceekineeciki miimiaki,* "all the passenger pigeons are dead," but in this case, it does imply extinction, because most of us are aware that the passenger pigeons were hunted to extinction in the early twentieth century. The animate intransitive verb stem *ceekinee-,* "all dead," requires an animate subject, such as in the earlier examples using ants and pigeons, but the context is important for the intended meaning if a speaker wants to imply extinction. There is no uniquely separate term in the Myaamia language that distinguishes death from extinction: verbal context extends the meaning of death to include extinction.

It is worth mentioning that the typical way to talk about the Myaamia language, in the language, is by using the verb stem *myaamiaatawee-.* When this is inflected with an indefinite subject marker, we get *myaami-*

aataweenki, which literally means "speaking the Myaamia language"—and this is the typical way one refers to the language. Because the notion of language is not expressed in a nominal form, applying a concept like *dead* to a language not only feels strange in discourse but poses grammatical problems with verb agreement. A concept like death can really only be applied to entities (animate or inanimate) and not to actions. Speaking a language is not a thing, and therefore it cannot die, but it is an action, and so a language can cease to be spoken. If one wants to express a lack of speakers of the Myaamia language, one would say *myaamiaataweeh-soona,* "no one speaks the Myaamia language," or would use a term like *poonaataweenki,* "the language ceases to be spoken," both of which are verbs. From a Myaamia philosophical perspective, a language cannot die, but it can stop being spoken. This is important, because there really isn't any notion of permanence implied here, and if the language is available in some other form, including audio or written recordings, then it's conceivable that it could be spoken again. This way of talking about a language renders both concepts *death* and *extinction* irrelevant to the topic of language loss and revitalization.

Maliseet-Passamaquoddy

Maliseet is an Eastern Algonquian language spoken along the St. John river valley in New Brunswick, Canada. The UNESCO interactive online atlas of endangered languages lists Maliseet as "severely endangered."[40] For UNESCO, "severely endangered" means that the "language is spoken by grandparents and older generations; while the parent generation may understand it, they do not speak it to children or among themselves."[41] The Summer Institute of Linguistics (SIL) lists Maliseet as "shifting," which is defined as follows: "The child-bearing generation can use the language among themselves, but it is not being transmitted to children."[42] SIL also provides the descriptive comment for *shifting* "middle-aged or elderly. Mildly positive attitudes. Increasing interest in the language in some places. English preferred by most youth."[43] National Geographic's *Enduring Voices* project provides an interactive map illustrating "language hot spots" where there are many languages clustered in a region in danger of going extinct. Sadly, Maliseet does not appear on the map.

These three popular linguistic and world heritage sites present three different assessments of the relative vitality of the Maliseet language. Whose expert analysis should be the guide for language advocates as they address perceived Maliseet language endangerment? What criteria and whose diagnoses are correct? How can communities decide on appropriate interventions when the experts cannot agree on the state of the Maliseet language? The lack of consensus among language experts need not be a constraint for Maliseet language advocates. The disagreement between experts is an opportunity to change the modes of analysis and diagnosis from endangerment and extinction. Instead of focusing on language endangerment, death, and extinction, community advocates can redirect their interventions toward community language life and emergent forms of language and cultural vitality.

The many expert pronouncements of Maliseet language endangerment present a trajectory that limits the options for language vitality. The biological organism metaphor is useful to a certain degree. When the metaphor blinds language experts from seeing innovative and creative uses of "endangered" languages, the rhetoric becomes a liability. For example, to use the Maliseet case, the UNESCO statement that Maliseet is "severely endangered" requires a logical calculus that promotes specific interventions such as documentation in the form of texts, audio/video recordings, descriptive analyses, and diverse modes of archiving. These are useful strategies for languages that have only a few speakers. However, Maliseet presents a case where there are speakers who are working on new forms of Maliseet interactions, such as an online dictionary, audio files of Maliseet stories, children's story books, television documentary programming, and so forth.[44] These efforts do produce the kinds of documentary artifacts of spoken Maliseet, but they also have the added benefits of creating new genres for the Maliseet language, making the Maliseet language relevant in contemporary settings and forming new social relations. Together, the benefits highlight the most significant aspect of rethinking expert rhetoric and practice. Documentary practices focus on language as a code that needs to be preserved. This renders language as a science object that can be taken out of context and dismembered into its constituent parts: phonemes, morphemes, syntactic structures, and semantic analyses. This strategy also ignores the collateral extinctions that accompany language extinction, such as "education, religion, knowledge, everyday social interactions, and identity."[45] In short, language extinction is a kind of "mass extinction" where particular ecologies—linguistic/ cultural/ecological—are at risk of disappearing. In contrast, the Maliseet language advocates recognize the value of the artifacts of spoken Maliseet as significant contributions to Maliseet language vitality, but the critical difference is that they are creating new domains for Maliseet language use. Rather than dismember the Maliseet language, these advocates look to reintegrate the many cultural resources that language mediates in an emergent Maliseet vitality.[46]

Maliseet emergent vitality is the commitment to integrating all social relations in everyday interactions to assure creative and innovative futures for language, culture, and identity while drawing from the foundational substrate of tradition and continuity. The documents that are created during activities—such as voice-overs for television programs, graphic novels, and the cartoon created for this chapter—are all the material artifacts of conversations between language advocates and speakers of each language community.[47] Emergent vitality places emphasis on the interaction between speakers who produce the texts rather than focusing on the texts at the expense of the social relations. The cartoon in this chapter is emergent vitality across three indigenous languages: Myaamia, Anishinaabemowin, and Maliseet. It is an example of creating new domains for indigenous languages, new social relations to produce the translations for the cartoon, and a creative use of heritage languages in the twenty-first century.

Anishinaabemowin

Anishinaabemowin is the language of a group who refer to themselves as the "People of the Three Fires"—the Odawa, Potawatomi, and Ojibwe— who migrated from the eastern Atlantic area to the Great Lakes watershed thousands of years ago. According to oral history, it was the language used by Nanaboozhoo; according to written records, it was one of several languages used during the fur trade era from the seventeenth to the nineteenth centuries, with multinational speakers of Anishinaabemowin in Montreal, Detroit, Chicago, Duluth, and along the mighty Michi-ziibing. It is currently used in more than two hundred Anishinaabe com-munities in Quebec, Ontario, Manitoba, Saskatchewan, Alberta, North Dakota, Michigan, Wisconsin, and Minnesota. This range of use is both a strength and a weakness. The subtle connectivity is not obvious to ex-ternal observers. For example, Anishinaabemowin is not one of the 2,467 languages mentioned in the UNESCO *Atlas of the World's Languages in Danger.*[48] The most common variant, Ojibwe, is noted in four locations as vulnerable, definitely endangered, and severely endangered. Odawa is not listed at all, and Potawatomi appears in only select locations in Canada, Kansas, Michigan, Oklahoma, and Wisconsin. This is not an indictment of international linguistic analysis. It is a lesson on invisibility. The current nations and states are the reason for the decline in the use of Anishinaabemowin, and as such, they obscure the dimensions and details of the language use. If one imagines instead the Great Lakes watershed as a base for the language and shades areas where it is now used less, the picture changes. If one traces river routes, feast locations, trees crooked as signposts, and trade sites, the language possesses a powerful past, one that echoes into the present in the names of many rivers, lakes, and cit-ies. Moving forward in time, one can also find a trace in the technology of chat rooms; in blogs, shared sites, and game platforms; and in visual and audio recordings. Anishinaabemowin has suffered and is changing but is not yet extinct. Certainly, like many other indigenous languages no longer used regularly in trade and education, it could be lost at any time between the generation unable or unwilling to teach and the generation unable or unwilling to learn.

Jaaginigeshkiwin/Wasteful Extravagance

One of the main reasons to teach Anishinaabemowin is to preserve aes-thetic diversity and encourage individual linguistic creativity. Students become speakers, singers, writers, and record keepers. As a language is learned, it becomes a part of a person's identity: an internal monologue is formed. With increasing use, the inner voice is shaped by Anishinaa-bemowin, and in some settings, at certain times, it is the primary expres-sion of the stream of consciousness. Even speakers of multiple languages will find there are certain thoughts and experiences tied to each language. Most importantly, in every speaker, there exists a new variation of the language the person shares with other speakers. No two speakers use the language in precisely the same way. Anishinaabemowin is a language that fosters complexity and variation. For instance, *Anishinaabe* is a both a verb that can be inflected in many ways and a morpheme that can be combined with other verbs. *Anishinaabe nind'aaw* (I am Anishinaabe) is related to *Nindanishinaabem* (I speak the Anishinaabe language), *Nindanishinaabemaaz* (I sing in the Anishinaabe language), and *Nindnishinaabebiiaan* (I write something in Anishinaabemowin). One could negate these sentences to indicate when these actions cease, but to say that the language no longer exists would require using the morphemes that most closely approximate extinction: *jaaginigeshki* (to be used up) and *angozo* (to disappear). Yet a construction that indicates the language has been "used up" or made to "disappear" does not imply the permanence of the word *extinction*. These verbs can change from something one does to something one does to another quite easily, and the "other" can be animate or inanimate. Verbs can also become nouns expressing wasteful extravagance, such as *jaaginigeshkiwin*. At the core of the language is a flexibility and denial of permanence that is so strong it serves as protection against extinction. The rules of construction provide a scaffold, and sometimes a part of that skeleton goes missing for a time, until it is found or a replacement is borrowed or invented. Linguists can work to determine the number and fluency of speakers, and languages do become silent. But until a lingometer or *Anishinaabidibabiigan* is created, we cannot know if the language is entirely lost. It may be in memories and landscapes waiting, like an unseen water table, to be replenished and remembered. Just as metaphors of extinction suggest that there is a "last speaker," there must then be a "first speaker" always on the edge of evolution. Perhaps chasing lost speakers has never been relevant and instead we should be nurturing the first speakers of the next version of the language.

Figure 9.3. Bernard Perley, "Unpacking Colonial Baggage in Native North America," panel 3 of 3, from the *Having Reservations* series. The complete cartoon appears in Appendix A. Translations of the cartoon into Myaamia, Maliseet, and Anishinaabemowin appear in Appendix B. Courtesy of the artist.

CONCLUSION: SURVIVING MASS EXTINCTION

Translating Invasive Concepts

The preceding cartoon presents the epistemological–ontological dilemma American Indian communities had to resolve as their worlds were transformed by invasive species and accompanying invasive concepts. Surviving the sixth extinction requires translating invasive concepts from 1492 to today. *Extinction* in the twenty-first century is the invasive concept the authors independently translated into their own heritage languages for the cartoon. The exercise was intended to compare their respective translations to see if they would vary in significant ways from each other and how much they differ from English. Discerning the range of differences across the three languages discussed in the chapter would indicate whether *extinction* and its English semantic properties reflected invasive concepts. The three languages provide the following translation for "it means 'extinction' has come to Native America!":

Myaamia
Ceekineeyoni kati pyaaki mihtohseenionkiši iišinaakwahki.
(It means that dying out has come to the place of the people.)

Maliseet
Nit ehta yakw, mehtapeksowakən peciyewiw skicinowihkohk!
([They're saying] "total population die-out" has arrived in Indian land!)

Anishinaabemowin
Anishinaabe-akiing gakina wii banaajichigaade.
(In Indian country all will be spent up/destroyed.)

Baldwin states that the Myaamia expression *ceekineeyoni* translates as "death to all" and "is grammatically a noun and was created to deal with the English concept of 'extinction' in the context of the cartoon. Since we were translating from the English and not semantically working from Myaamia, we used a noun form which is typical when dealing with difficult translations."[49] The Maliseet translation of *mehtapeksowakon* is also a noun form that describes population die-out but does not imply total species extermination. The "die-out" can be a seasonal occurrence, but the expectation is that the population will return the following season. However, Henrietta Black explains that the free translation in the context of the cartoon "sounds worse in Maliseet: 'We are your exterminators!'"[50] Finally, Noodin explained why she used *banaajichigaade* instead of *jaaginigeshkiwin* for the cartoon: "Neither word perfectly implied all of what extinction means. I think *jaaginigeshkiwin* is closer to the scientific implications but the cartoon used *binaajichigaade* because it alludes to intentional destruction."[51] The translations of *extinction* in all three languages reveal the periphrasis required to approximate English semantic properties that extinction in the twenty-first century conveys. The authors are English speakers and understand the semantic properties of *extinction*. But they also have access to their heritage languages and recognize that some English concepts can have unintended but detrimental consequences for American Indian vitalities.[52] This critical awareness of the differences and implications that invasive concepts have for American Indian communities is a key factor for surviving the sixth extinction.

The Semantics of Survival

If settler societies want to know what surviving mass extinctions will be like after species extinctions owing to human-induced environmental changes, then all they need to do is look at American Indian communities. When American Indian communities in North America (in the United States in particular) reached their nadir at the end of the nineteenth century, many of the settler scholars and advocates were ready to write off American Indians as tragic victims of progress and civilization. Today, the descendants of those settler societies are anguishing over their own sense of mortality in the face of climate change and species extinctions. The "curse of unimaginable magnitude on future generations" was unleashed on American Indians by settler societies as they colonized the Western Hemisphere.[53] In North America, collectively, the invasive species, invasive concepts, and changes in the landscape were the first wave of mass extinctions in our collective sixth extinction crisis. There is documentary evidence that records the day our sixth extinction began: October 12, 1492. The excerpt from Columbus's journal of the first voyage reveals the importation of invasive concepts as he denied the native peoples their sovereignty ("they ought to make good slaves"), their religion ("they had no religion of their own"), and their languages ("will bring . . . them to Your Highnesses so that they may learn to speak"). Not only did the explorers and settlers bring material objects with their baggage but they also brought their ideas and concepts. Among those concepts were Christianity, greed, war, conquest, and self-serving superiority. The problem with invasive concepts is that the invaders do not realize that their beliefs and knowledge systems have evolved in particular contexts and environments, and when they step foot in a new world, their perceptions and experiences do not prepare them to see the world the same way the native peoples do. Columbus was just the first to initiate processes that would propel the human species toward the catastrophe of mass extinction. The first convulsion of the sixth extinction obliterated entire tribes, languages, cultures, ecosystems, species, and landscapes. The extent of population loss, species loss, and habitat loss will never be known, but we do know that some American Indian communities endured and survived more than five hundred years of dramatic transformations in the opening act of human extinction. Can we survive the next convulsion?

The authors shared their personal experiences and insights into surviving mass extinction through their work in repatriating American Indian linguistic sovereignty. Baldwin dared to challenge expert opinion and refused to accept that the Myaamia language was extinct. He made the deliberate decision to change the metaphor from *extinct* to *asleep*. Doing so allowed the collaboration with linguist David Costa, the Miami Tribe of Oklahoma, and Miami University in their efforts to "awaken" the Myaamia. Perley realized that expert diagnostic practices that focus on endangerment and extinction are metaphors of impending and final death. Rather than focus on language death, Perley argues that language advocates and experts should focus on language life. The creation of new domains of language use is the result of developing new social relations. Those new relations and their concomitant language domains are the kinds of emergent vitality that will assure the continued relevance of Maliseet in community lives. Noodin asserts a reconfiguration of sovereignty that ignores settler-imposed boundaries and categories. The recognition that the Great Lakes watershed represents Anishinaabe sovereign space is also recognition of the complex biome that is reflected in the sounds, words, and expressions of Anishinaabemowin. All three share the experience of surviving mass extinctions, but they also share a vision that realigns epistemological stances between settler populations and indigenous populations. Such an alignment may not forestall the looming extinctions, but they do offer perspectives for how American Indian communities survived the first convulsion of our sixth extinction. Perhaps the "curse of unimaginable magnitude on future generations" cannot be broken. The examples, Myaamia, Maliseet, and Anishinaabe, are only three of many American Indian communities who have survived the first round of human extinction. As we all anticipate and face the next round of mass extinctions, we can look to American Indian strategies of awakening, emergent vitality, and sovereignty for guidance in surviving what will become a global New World.

APPENDIX A

Figure 9.4. Bernard Perley, "Unpacking Colonial Baggage in Native North America," from the *Having Reservations* series. Translations of the cartoon into Myaamia, Maliseet, and Anishinaabemowin appear in Appendix B. Courtesy of the artist.

APPENDIX B

The following are the transcripts of the translations of the cartoon into Myaamia, Maliseet, and Anishinaabemowin.

Myaamia

(With assistance from David Costa)

1.a. Are you lost?
Waawiiyiikwi-nko
(Have you lost your way?)

1.b. Are you visiting?
Kiiwikawiyiikwi-nko
(Are you visiting?)

1.c. Neither. We are here to stay.
Moohci. Oowaaha kati wiiyaahkiaanki
(No. Here is where we will stay.)

1.d. And we brought lots of baggage.
Neehi wiihsa piitwaanki nintaantaleminaana.
(And we brought a lot of our things.)

2.a. Christianity.
Alaaminaayoni.
(Christian prayer.)

2.b. War.
Mihšihkatwi.

2.c. Greed.
Myaalhkaayoni.

2.d. Diseases.
Mintaayoni.
(Sickness.)

2.e. Conquest.
Weewiinkamihiweenki.
(One defeats people.)

3.a. What do you think this means?
Taani ooniini iišinaakwahki?

(What does this mean?)

3.b. It means "extinction" has come to Native America!
Ceekineeyoni kati pyaaki mihtohseenionkiši iišinaakwahki.
(It means that dying out has come to the place of the people.)

Maliseet

Having Reservations
Mate Sapitahamawiwa

(They [I] don't trust them.)
Unpacking Colonial Baggage.
Nməni Pihtkəmatmənenol Psiw Kekwil Peciptowekw.
(Taking out what items we brought.)

1.a. Are you lost?
Kəhskahapa?
(Are you two lost?)

1.b. Are you visiting?
Kpeciwikwamkepa?
(Have the two of you come to visit [our place of dwelling]?)

1.c. Neither. We're here to stay.
Kahte. Ntahskəmi petotepon nilon.
(No. We'll stay forever.)

1.d. And we brought lots of baggage!
Naka milikən kisi petotəhktowekw.
(We moved here with everything.)

2.a. Christianity.
Papahtəmwakən.
(Religion.)

2.b. War.
Matnətimək.
(Fighting each other.)

2.c. Greed.
Akwamhotikewakən.
(Getting more than you need/have ["you want it all"].)

2.d. Diseases.

Khsinokewakənəl.
(All kinds of illnesses.)
2.e. Conquest.
Piyemhanhtowiktəmakən.
(Overpowering others[?].)
3.a. What do you think this means?
Kekw nət klitahas ktiokonok?
(What do you think he's telling us?)
3.b. It means "extinction" has come to native America!
Nit ehta yakw, mehtapeksowakən peciyewiw skicinowihkohk!
([They're saying] "total population die-out" has arrived in Indian land.)
["It sounds worse in Maliseet: 'We are your exterminators!'"—
Henrietta Black]

Anishinaabemowin

Having Reservations
Ji-ayamaang Ishkoniganan
(We are having leftovers/reservations.)

Unpacking colonial baggage in native North America.
Zhaabwaabanjigaazowag dezhiikejig ge gegeti-zhewebiziwaad Anishi-naabe-akiing.
(They are seen through, the settlers, as truly-they behave in native North America.)

1.a. Are you lost?
Gi wanishinim ina?
1.b. Are you visiting?
Gi mawadishiwem ina?
1.c. Neither. We're here to stay.
Gaawiin. Ni wii dazhiikemin omaa.
(We [just us] will stay/settle here.)
1.d. And we brought lots of baggage!
Miinawaa in gii biidoomin anoonji indo'zhewebiziwiniminaanan!
(And we brought many of our behaviors.)

2.a. Christianity.
Wenzhishimanidookewin.
(Making holy of one specific spirit.)

2.b. War.
Miigaadiwin.
(Physically fighting each other.)

2.c. Diseases.
Majiwaapinewinan.
(Badness that happens in the body.)

2.d. Greed.
Mindaweshkiwin.
(Habitual greed.)

2.e. Conquest.
Mamaazhiwin.
(Note: *zhaagoodin* [overthrow someone] or *mamaazhi* [to do badness to someone].)

3.a. What do you think this means?
Aaniin enendaman ge-edamowaad?
(What do you think they mean [by what they say]?)

3.b. ???
Ni manji iidog.
(That's unknown really.)

3.c. It means "extinction" has come to native America.
Anishinaabe-akiing gakina wii banaajichigaade.
(In Indian country all will be spent up/destroyed.)

NOTES

The authors want to express their thanks and great appreciation to David Costa for his contributions to consultation on translation of the cartoon from English to Myaamia; to Henrietta Black for her contributions to consultation on translation of the cartoon from English to Maliseet; and to Michael Zimmerman for his contributions to consultation on translation of the cartoon from English to Anishinaabemowin. This collaborative effort is another example of indigenous language life as part of the larger social relations among diverse community conversations.

1. For the popularization of the concept of the sixth extinction in trade publications, see, e.g., Elizabeth Kolbert, *The Sixth Extinction: An Unnatural History* (New York: Henry Holt, 2014); in academic symposia, see UW–Milwaukee's Center for 21st Century Studies' After Extinction conference (spring 2015), where participants outlined many of the themes that resonated across disciplines and popular media, http://www4.uwm.edu/c21/conferences/2015afterextinction /afterextinction.html; and in popular media, see "Which Species Will Survive the Earth's Sixth Mass Extinction?," *The Conversation* (blog), September 22, 2015, http://theconversation.com/which-species-will-survive-the-earths-sixth -mass-extinction-47893.

2. Richard Leakey and Roger Lewin, *The Sixth Extinction: Biodiversity and Its Survival* (London: Widenfeld and Nicholson, 1996), 224.

3. Marc Abélès, *The Politics of Survival* (Durham, N.C.: Duke University Press, 2010); Ulrich Beck, *Ecological Politics in an Age of Risk* (Cambridge: Polity Press, 1995).

4. Daniel Nettle and Suzanne Romaine, *Vanishing Voices: The Extinction of the World's Languages* (Oxford: Oxford University Press, 2000); K. David Harrison, *The Last Speakers: The Quest to Save the World's Most Endangered Languages* (Washington, D.C.: National Geographic, 2010); Nicholas Evans, *Dying Words: Endangered Languages and What They Have to Tell Us* (Malden, Mass.: Wiley-Blackwell, 2010).

5. See K. David Harrison, *When Languages Die: The Extinction of the World's Languages and the Erosion of Human Knowledge* (Oxford: Oxford University Press, 2007), for his forceful argument on the loss of human knowledge; Luisa Maffi's argument for biocultural diversity loss in her book *On Biocultural Diversity: Linking Language, Knowledge, and the Environment* (Washington, D.C.: Smithsonian Institution Press, 2001); and Peter Mühlhäusler, *Linguistic Ecology: Language Change and Linguistic Imperialism in the Pacific Region* (London: Routledge, 1996).

6. B. W. Ife, *Christopher Columbus: Journal of the First Voyage* (Warminster, U.K.: Aris and Phillips, 1990), 29.

7. Ibid.

8. Ibid., 31.

9. See the highly collaborative and community engaged video *Myaamiaki*

Eemamwiciki: Miami Awakening. Sandy Sunrising Osawa, Yasu Osawa, Daryl Baldwin, David J. Costa, and Miami Tribe of Oklahoma, *Myaamiaki Eemamwiciki: Miami Awakening* (Seattle, Wash.: Upstream Productions and Miami Tribe of Oklahoma, 2008), DVD.

10. Gary F. Simmons and M. Paul Lewis, "The World's Languages in Crisis: A 20-year Update," in *Responses to Language Endangerment: In Honor of Mickey Noonan,* ed. Elena Mihas, Bernard Perley, Gabriel Rei-Doval, and Kathleen Wheatley (Amsterdam: John Benjamins, 2013), 13.

11. Ibid., 16.

12. William Cronon, *Changes in the Land: Indians, Colonists, and the Ecology of New England* (New York: Hill and Wang, 1983).

13. Steven M. Stanley, *Extinction* (New York: Scientific American Library, 1987), ix.

14. Ibid., x.

15. Bernard C. Perley, "Last Words, Final Thoughts: Collateral Extinctions in Maliseet Language Death," in *The Anthropology of Extinction: Essays on Culture and Species Death,* ed. Genese Marie Sodikoff, 127–42 (Bloomington: Indiana University Press, 2012).

16. Ibid., 133.

17. Harrison, *When Languages Die*; Lenore A. Grenoble, "Endangered Languages," in *Living, Endangered, and Lost: One Thousand Languages,* ed. Peter K. Austin, 216–35 (Berkeley: University of California Press, 2008); Simmons and Lewis, "The World's Languages in Crisis."

18. Daniel Nettle and Suzanne Romaine, *Vanishing Voices: The Extinction of the World's Languages* (Oxford: Oxford University Press, 2000); Harrison, *When Languages Die.*

19. Perley, "Last Words, Final Thoughts"; Bernard Perley, "Remembering Ancestral Voices: Emergent Vitalities and the Future of Indigenous Languages," in *Responses to Language Endangerment,* 243–70.

20. Russell Thornton, "Historical Demography," in *A Companion to the Anthropology of American Indians* (Malden, Mass.: Blackwell, 2004), 29.

21. Colin G. Calloway, *First Peoples: A Documentary Survey of American Indian History,* 2nd ed. (Boston: Bedford/St. Martins, 2004), 378.

22. Duncan Campbell Scott, deputy superintendent general of Indian affairs, testimony before the Special Committee of the House of Commons examining the Indian Act amendments of 1920, National Archives of Canada, Record Group 10, vol. 6810, file 470-2-3, vol. 7, 55 (L-3) and 63 (N-3); available through the Critical Thinking Consortium, "Source Docs," http://tc2.ca/sourcedocs /uploads/images/HD%20Sources%20(text%20thumbs)/Aboriginal%20History /Residential%20Schools/Residential-Schools%2010.pdf.

23. "'Vanishment'—irrevocable, complete, irreversible—is the first axiom of revisionism." A. J. Prats, *Invisible Natives: Myth and Identity in the American Western* (Ithaca, N.Y.: Cornell University Press, 2002), 127. See also Gerald

Vizenor, *Fugitive Poses* (Lincoln: University of Nebraska Press, 1998), and "Native American Traditions," Pluralism Project, http://www.pluralism.org/religion/native-american.

24. Quoted in Robert A. Trennert, "The Business of Indian Removal: Deporting the Potawatomi from Wisconsin, 1851," *The Wisconsin Magazine of History* 63, no. 1 (1979): 46, http://www.jstor.org/stable/4635374. Trennert's note cites the following source: Alexis Coquillard to G. W. Ewing, July 16, August 11 and 17, 1851, in the Ewing Papers; *Milwaukee Daily Free Democrat,* August 2, 1851.

25. Simon Pokagon, *Ogimawkwe Mitigwaki: A Novel* (East Lansing: Michigan State University Press, 2011), 83.

26. Prats, *Invisible Natives,* 127.

27. This was penned by John Brown Dillon in 1848 as a commentary (excuse-making session) for why the Miamis were removed and why they were destined to vanish. Dillon, "The National Decline of the Miami Indians," *Proceedings of the Indiana Historical Society, 1830–1886* 1, no. 4 (1897): 141. "John Brown Dillon (1807?–1879) was the editor of Logansport, Cass County Canal-Telegraph (1834–1842), the Indiana State librarian (1845–1851), the secretary for the State Board of Agriculture (1850s), a clerk in the U.S. Department of the Interior (1863–1871), the House Military Affairs Committee (1871–1875), secretary of the Indiana Historical Society (1859–1879), and author of the *History of Indiana* (1843) and other works on the history of Indiana, Indians and the Old Northwest." "John Brown Dillon," Finding Aid Index, Indiana State Library, http://www.in.gov/library/finding-aid/3918.htm.

28. John Dyneley Prince, introduction to *Kulóskap the Master: And Other Algonkin Poems,* trans. Charles Godfrey Leland and John Dyneley Prince (New York: Funk and Wagnalls, 1902), 40.

29. *Merriam-Webster Online,* s.v. "extinction," http://www.merriam-webster.com/dictionary/extinction.

30. See *Myaamiaki Eemamwiciki: Miami Awakening.*

31. Daryl Baldwin, personal communication with Bernard Perley, June 28, 2015.

32. Brian D. McInnes, *Sounding Thunder: The Stories of Francis Pegahmagabow* (East Lansing: Michigan State University Press, 2016), 178.

33. Andrew Medler, *Weshki-Bmaadzijig ji-noondmowaad: "That the Young Might Hear": The Stories of Andrew Medler as Recorded by Leonard Bloomfield,* ed. J. Randolph Valentine (London, Ont.: Centre for Research and Teaching of Canadian Native Languages, 1998), 127.

34. Sam Osawamick, *Stories of Sam Osawamick from the Odawa Language Project,* ed. G. L. Piggott, Edna Manitowabi, and John Nichols (Winnipeg: University of Manitoba, 1985), 12.

35. Margaret Noodin, personal communication with Bernard Perley, November 10, 2015.

36. Henrietta Black, personal communication with Bernard Perley, July 5, 2015.

37. See also Wesley Y. Leonard, "When Is an Extinct Language Not Extinct? Miami a Formerly Sleeping Language," in *Sustaining Linguistic Diversity: Endangered and Minority Languages and Language Varieties,* ed. Kendell A. King, Natalie Schilling-Estes, Lyn Fogle, Jia Jackie Lou, and Barbara Soukup, 23–33 (Washington, D.C.: Georgetown University Press, 2008).

38. Daryl Baldwin and Julie Olds, "Miami Indian Language and Cultural Research at Miami University," in *Beyond Red Power: American Indian Politics and Activism since 1900,* ed. Daniel M. Cobb and Loretta Fowler, 280–90 (Santa Fe, N.M.: School of Advanced Research, 2007).

39. Christopher Moseley, ed., *Atlas of the World's Languages in Danger,* 3rd ed. (Paris: UNESCO Publishing, 2010), http://www.unesco.org/languages-atlas /en/atlasmap.html.

40. Ibid.

41. "Endangered Languages," in Moseley, *Atlas of the World's Languages in Danger,* http://www.unesco.org/new/en/culture/themes/endangered-languages /atlas-of-languages-in-danger/.

42. M. Paul Lewis, Gary F. Simons, and Charles D. Fennig, eds., "Language Status," in *Ethnologue: Languages of the World,* 19th ed. (Dallas, Tex.: SIL International, 2016), http://www.ethnologue.com/about/language-status.

43. "Malecite-Passamaquoddy," ibid., http://www.ethnologue.com/language /pqm.

44. For the online dictionary, see Passamaquoddy-Maliseet Language Portal, http://pmportal.org/; for audio files of Maliseet stories, see Koluskap: Stories from Wolastoqiyik, http://website.nbm-mnb.ca/Koluskap/English/Stories /StoryList.php; for children's story books, see Say It First: Aboriginal Language Revitalization, http://www.sayitfirst.ca/projects/children-books; for television documentary programming, see Samaqan Water Stories, http://www.samaqan .ca/.

45. Perley, "Last Words, Final Thoughts," 133.

46. Bernard C. Perley, "Language as an Integrated Cultural Resource," in *A Companion to Cultural Resource Management,* ed. Thomas F. King, 203–20 (Malden, Mass.: Blackwell, 2011); Perley, "Remembering Ancestral Voices."

47. For examples of television programs, see Perley, "Remembering Ancestral Voices," 256–57; for examples of graphic novels, see ibid., 258–60.

48. Moseley, *Atlas of the World's Languages in Danger.*

49. Daryl Baldwin, personal communication with Bernard Perley, November 8, 2015.

50. Henrietta Black, personal communication with Bernard Perley, November 10, 2015.

51. Margaret Noodin, personal communication with Bernard Perley, November 10, 2015.

52. See Appendix B for additional translations of the cartoon. For example, the title of the series is *Having Reservations.* The title was intended to evoke the American Indian experience in two ways. First, many communities were forced to live on lands set aside for their residence after government relocation. Hence the term *reservations* became a common state of experience. Second, the title reflects the pervasive doubts that American Indians have regarding any colonial or governmental statement or promise. The authors provided translations for *Having Reservations,* but as Baldwin notes, "puns are nearly impossible to translate because of the semantics . . . so I will propose this: *wanimihsoolamoci* (let him not fool us)." Daryl Baldwin, personal communication with Bernard Perley, November 10, 2015. Baldwin's translation is very similar to the Maliseet: *Mate Sapitahamawiwa* (they [I] do not trust them). However, Noodin takes a different approach to the pun and provides *Ji-ayamaang Ishkoniganan* (we are having leftovers/reservations).

53. Leakey and Lewin, *The Sixth Extinction,* 224.

Acknowledgments

This volume, like the two preceding it, is based on one of the annual spring conferences of the Center for 21st Century Studies (C21) at the University of Wisconsin–Milwaukee (UWM). The topic of the 2015 conference (and this volume), "After Extinction," grew out of C21's annual theme, Humanities Futures, which was chosen to support the declaration of 2014–15 as the Year of the Humanities at UWM, a designation that turned out to be more honorific than fiscal. In deciding on the topic of "After Extinction," we were especially pleased that it could be taken as an ironic comment on Humanities Futures, as suggesting the possibility that the humanities themselves could be headed to a future of extinction.

Little did we know that in the course of that academic year, it would be not only the humanities but public higher education more generally that looked to be threatened with extinction. While we had all been aware of the declining levels of state support for publicly funded higher education over the past few decades, it really was only during the 2014–15 academic year that one could begin seriously to imagine the extinction of public universities in America. In Louisiana, the university system was being forced by a recalcitrant state legislature to make preparations for bankruptcy. In Illinois, a Republican legislator had proposed legislation to privatize all Illinois public universities in six years. And here in Wisconsin, Governor Scott Walker threatened to eviscerate the Wisconsin Idea's pursuit of truth, cut $250 million from the University of Wisconsin System, and emboldened UW System president Ray Cross and the Board of Regents to dramatically weaken faculty governance, tenure, and academic freedom. Now, in 2017, as this volume is entering production, Kansas and Missouri are debating proposals to eliminate tenure, and states like Iowa, Illinois, and others continue to suffer draconian budget cuts. It is no longer unthinkable to imagine that we may soon have to consider what the public university will look like after extinction.

In organizing the After Extinction conference, I was joined by postdoctoral fellow Gloria Kim and deputy director Emily Clark. The organizing troika was supported with good cheer and better skill by associate director John Blum, business manager Annette Hess, and our two graduate project assistants, Kayla Payne and Nick Proferes; Nick's outstanding design work on the conference website and program is especially worth noting. Finally, the editing and production of this book could not have happened without the extraordinary work of John Blum, who serves as C21's editor. Over the course of my term as C21 director, I benefited immensely from John's counsel and comradeship; this volume benefited from his skills as an editor and from his wealth of experience in manuscript preparation. I would also like to thank Doug Armato, director of the University of Minnesota Press, for his role in bringing the C21's book series to Minnesota, and Danielle Kasprzak, humanities editor at Minnesota. Finally, I thank Anne Carter. Despite the incredible team of collaborators on this project, responsibility for any and all errors is, as always, mine.

Daryl Baldwin is director of the Myaamia Center at Miami University in Oxford, Ohio. He is coauthor (with David J. Costa) of *Myaamia Neehi Peewaalia Kaloosioni Mahsinaakani: A Miami-Peoria Dictionary.*

Claire Colebrook is the Edwin Erle Sparks Professor of English at Pennsylvania State University. She is the author of *Death of the PostHuman: Essays on Extinction, Volume 1; Sex after Life: Essays on Extinction, Volume 2;* and *Blake, Deleuzian Aesthetics, and the Digital.*

William E. Connolly is the Krieger–Eisenhower Professor of Political Science at Johns Hopkins University. He is the author of *Aspirational Fascism: The Struggle for Multifaceted Democracy under Trumpism* (Minnesota, 2017); *Facing the Planetary: Entangled Humanism and the Politics of Swarming; The Fragility of Things: Self-Organizing Processes, Neoliberal Fantasies, and Democratic Activism; A World of Becoming;* and *Capitalism and Christianity, American Style.*

Ashley Dawson is professor of English at the CUNY Graduate Center. He is the author of *Extinction: A Radical History; The Routledge Concise History of Twentieth-Century British Literature;* and *Mongrel Nation: Diasporic Culture and the Making of Postcolonial Britain.*

Richard Grusin is professor of English and director of the Center for 21st Century Studies at the University of Wisconsin–Milwaukee. He is author of *Premediation: Affect and Mediality after 9/11; Culture, Technology, and the Creation of America's National Parks;* and (with Jay David Bolter) *Remediation: Understanding New Media.*

Joseph Masco is professor of anthropology at the University of Chicago. He is the author of *The Theater of Operations: National Security Affect from the Cold War to the War on Terror* and *The Nuclear Borderlands: The Manhattan Project in Post–Cold War New Mexico.*

Nicholas Mirzoeff is professor of media, culture, and communication at New York University. He is the author of *How to See the World* and *The Right to Look: A Counterhistory of Visuality* and the editor of *The Visual Culture Reader.*

Margaret Noodin is professor of English at the University of Wisconsin–Milwaukee. She is the author of *Bawaajimo: A Dialect of Dreams in Anishinaabe Language and Literature* and *Weweni: Poems in Anishinaabemowin and English.*

Jussi Parikka is professor of technological culture and aesthetics at the Winchester School of Art, University of Southampton. He is the author of several books on media theory and digital culture, including *Digital Contagions*; *A Geology of Media* (Minnesota, 2015); *What Is Media Archaeology?*; and *Insect Media* (Minnesota, 2010).

Bernard C. Perley is professor of anthropology at the University of Wisconsin–Milwaukee. He is the author of *Defying Maliseet Language Death: Emergent Vitalities of Language, Culture, and Identity in Eastern Canada* and coeditor (with Elena Mihas, Gabriel Rei-Doval, and Kathleen Wheatley) of *Responses to Language Endangerment: In Honor of Mickey Noonan.*

Cary Wolfe is the Bruce and Elizabeth Dunlevie Professor of English at Rice University. He is the author of *Before the Law: Humans and Other Animals in a Biopolitical Frame*; *What Is Posthumanism?* (Minnesota, 2009); and *Animal Rites: American Culture, the Discourse of Species, and the Posthumanist Theory.*

Joanna Zylinska is professor of new media and communications at Goldsmiths, University of London. She is author of *Minimal Ethics for the Anthropocene* and *Bioethics in the Age of New Media* and is the translator of Stanisław Lem's *Summa Technologiae* (Minnesota, 2014).

Index

petrochemicals: and capitalism, 77, 94–95, 97, 101, 179; dangers of, 71, 80; and film stock, 97. *See also* fossil fuels

Pitawanakwat, Alphonse, 212

planetary boundaries concept, 90–92, *91*

Pokagon, Simon, 209

posthistory, xii–xiii, 29–31, 38–39, 42–43, 45n2

premediation, x–xi, 29, 48n43

racism, 4–5, 19, 22n5; and Anthropocene, xv, 123–28, 130, 138, 142–43; and extinction, xi, xv, xvi; scientific, 134; systemic, 123–24. *See also* white supremacy

reductionism, 3–4, 22n5, 40

Revive and Restore, 178, 181, 183. *See also* de-extinction

Schmitt, Carl, 34

Serres, Michel, 27, 46n21

Sharma, Sarah, xii, 38, 41

Shaviro, Steven, 30

Singer, Peter, 154–55, 158–60, 163

singularity, 31, 39, 46n14, 47n30

Six, Nicole, and Paul Petritsch: *Spatial Intervention I,* 97–100, *98, 99*

16mm (Mangrané), 95–97, *96*

sixth extinction, vii–xi, xvi, 87–88, 201–3; imagining the, 28, 52–53, 161; responses to, 174–77; skepticism about, 186; surviving the, 220–23; and utilitarianism, 154–56, 160

Sixth Extinction: An Unnatural History, The. See Kolbert, Elizabeth

sixth mass extinction. *See* sixth extinction

Skaer, Lucy: *Leviathan's Edge,* 85–88, *85, 86*

Snæbjörnsdóttir, Bryndís, and Mark Wilson: *Trout Fishing in America and Other stories,* xiv, 109, *111, 119*

sociocentrism, 1–3, 5, 10, 12, 14

Spatial Intervention I (Six and Petritsch), 97–100, *98, 99*

spirituality, 18, 20

Stoermer, Eugene, 75

Sugimoto, Hiroshi, 52, 69n31; *Lost Human Genetic Archive,* xii–xiii, 58–59, *59*

Suicide Narcissus (Walker), xiii, 83–85, 96–97, 98–99, 100, 102. *See also* individual artists in the art exhibition: Bauman, Thomas; Epaminonda, Haris, and Daniel Gustave Cramer; Mangrané, Daniel Steegman; Paterson, Katie; Six, Nicole, and Paul Petritsch; Skaer, Lucy

Suns (from Sunsets) from Flickr (Umbrico), xiii, 62–63, *63*

Talbot, W. H. Fox, 55, 68n14, 134

Tau Sling (Bauman), 92–95, *93*

Taylor, Charles, 10, 25n17

traces: fossil, 87, 134; historical, 29; human, 96; material, 117, 120; memory, 15, 52; technological, 218

Trout Fishing in America and Other stories (Snæbjörnsdóttir and Wilson), xiv, 109, *111, 119*

Tsing, Anna, 6

Umbrico, Penelope, 52, 65; *Suns (from Sunsets) from Flickr,* xiii, 62–63, *63*

UNESCO: *Atlas of the World's Languages in Danger,* xviii, 215–16, 218

utilitarianism, 152–56, 159–60, 163–69; and Kantianism, 10